Turn Right At Orion

OTHER BOOKS BY
MITCHELL BEGELMAN

Gravity's Fatal Attraction (with Sir Martin Rees)

Turn Right At Orion

Mitchell Begelman

Helix Books

Perseus Publishing
Cambridge, Massachusetts

A CIP record for this book is available from the Library of Congress
ISBN: 0–7382–0207-X

Perseus Publishing is a member of the Perseus Books Group

Text design by Jeff Williams
Set in 10-point Sabon by the Perseus Books Group

1 2 3 4 5 6 7 8 9 10—03 02 01 00
First printing, August 2000

Find us on the World Wide Web at http://www.perseusbooks.com

To my parents

Contents

Introduction

Translator's Note

I came across the following memoir under distressing circumstances: at an auction of surplus documents to raise funds for a new storage wing at the Global Library. The thought of this valuable information being sold to the highest bidder is appalling, but such are the times we live in. The author appears to come from a well-studied planet in a galaxy about 60 million light-years away, known to its erstwhile occupants as "Earth." If we are correct in our attribution, he belongs to a civilization that warrants more than a passing mention in our Core Curriculum because of the early and curious way in which its technology developed. What appears to clinch the identification is the author's reference, in the very first paragraph, to a scientist known (in his language) as Kepler. Of course, information on this pivotal figure was eventually disseminated cosmos-wide in Earth's copious (though brief) transmissions, but the subtle "ownership" expressed by the author suggests that he and Kepler have common cultural origins. There are also linguistic clues, turns of phrase that seldom show up in adaptations of Earth's language by neighboring civilizations. Fortunately, our records of their transmissions have enabled us to produce what we feel must be an accurate reconstruction of their language; this forms the basis for my translation.

Several things are remarkable about this document. The author's mixture of experiential metaphor with scientific theory is highly unusual; most records we possess of Earth's achievements in astrophysics indicate a greater detachment on the part of the observer. The science is generally of high quality, with understandable lapses. Notably, there are significant gaps in the author's comprehension of what causes stars to explode, his grasp of pulsar physics leaves much to be desired, and his interpretation of how jets of plasma escape from the vicinity of black holes is sketchy at best. But the glimpses we are granted into this exotic being's direct responses to cosmic phenomena more than compensate for these quibbles. Although the narrator seems curiously reserved—perhaps because of his separation from his home planet—the emotional reactions he does express are strikingly similar to those recorded by our own early explorers. We are swept up in his compulsion, even though the initial reasons for his journey (and his journal) are never satisfactorily explained. Solitary space travel has never been the norm in our culture, and we are at a loss to interpret the list of reasons given. Do we take them at face value? Or is the author unwilling or unable to divulge his real motivations? Is he, perhaps, an exile from his home planet? Or a scout, for a planned mission of colonization or emigration, who has "gone native"?

Whatever his motivations, the narrator displays an admirable command of technology. He recognizes that the only effective method of space travel involves motion at very close to the speed of light. He is adept at manipulating this motion to maintain the structural integrity of both his craft and himself, even when he finds he has misjudged the harsh conditions of space. We find frequent evidence of his resourcefulness, such as his use of hibernation to circumvent the limitations imposed by his species's brief life span.

In *Turn Right at Orion*, we are treated to a journey of great scope, far longer than initially planned, and are privy to the narrator's discoveries and dilemmas as he travels farther and farther afield. In fact, some passages (descriptions of the jet and extended galactic halo, in particular) actually suggest that he was in the center of *our* galaxy when he transmitted this memoir.

The title of the memoir was chosen by its author and is somewhat problematical. We understand that "Orion" refers to a region of vigorous star and planet formation—long since dissipated—that was located about 1500 light-years from the author's home planet at the time of his departure. The author's visit to this region, recounted in the lengthy fourth section of the memoir, clearly made a strong impression on him. I regard it as the turning point in his story. But we have absolutely no idea what is meant here by the phrase *turn right*; such a phrase makes little sense in the context of cosmic navigation. We can only surmise that it is a colloquialism or an obscure cultural reference.

Whereas in more prosperous (and enlightened) times the publication of such a document without exhaustive footnotes and a gloss would have been unthinkable, there is little support for such scholarly effort today. However, to discover and rescue such a prize and *not* to share it, in whatever form, would seem an even greater outrage. I offer my translation of this manuscript unadorned. Make of it what you will.

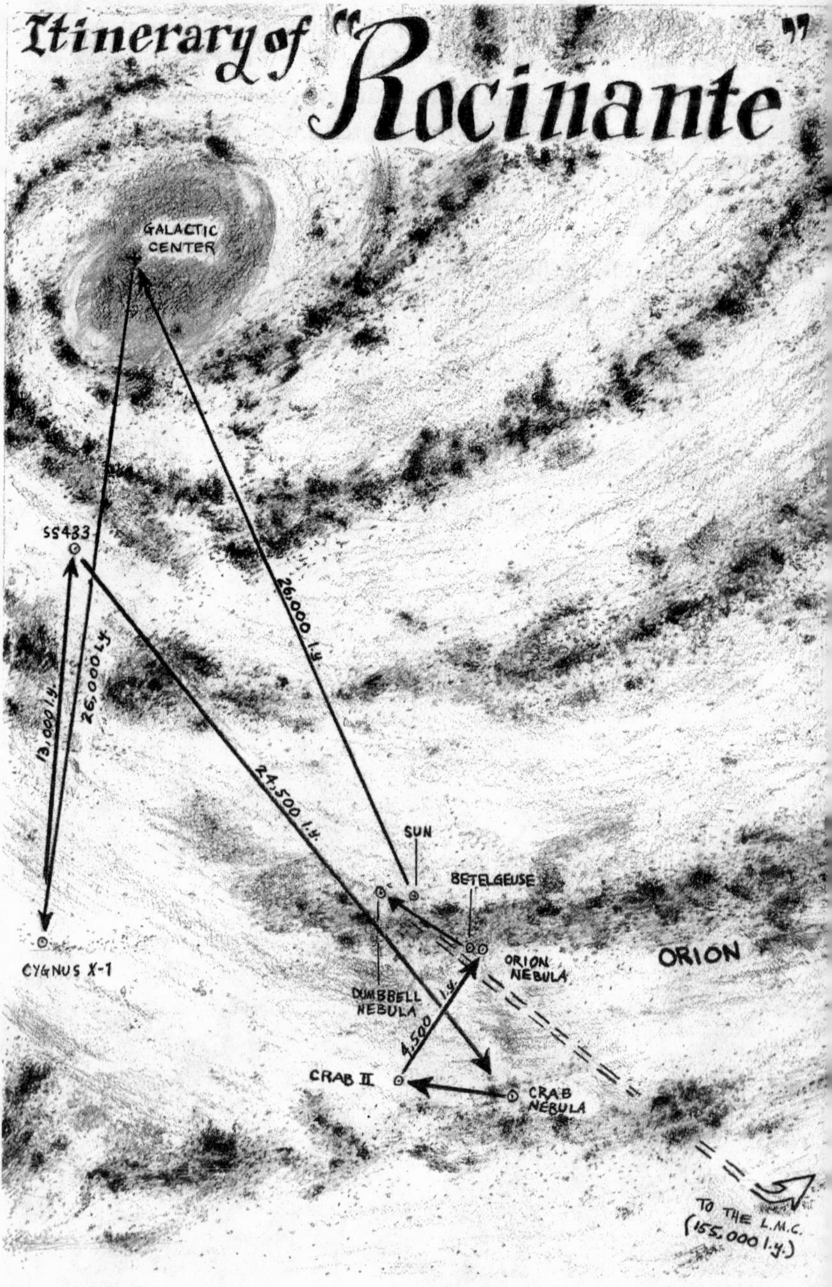

Itinerary of "Rocinante"

GALACTIC CENTER

SS433

26,000 l.y.

13,000 l.y.

26,000 l.y.

24,500 l.y.

SUN

BETELGEUSE

CYGNUS X-1

ORION NEBULA

ORION

DUMBBELL NEBULA

4,500 l.y.

CRAB II

CRAB NEBULA

TO THE L.M.C. (155,000 l.y.)

The Local Group

TO VIRGO CLUSTER (60,000,000 l.y.)
THE MILKY WAY
LARGE MAGELLANIC CLOUD
SMALL MAGELLANIC CLOUD
ANDROMEDA GALAXY

ARM

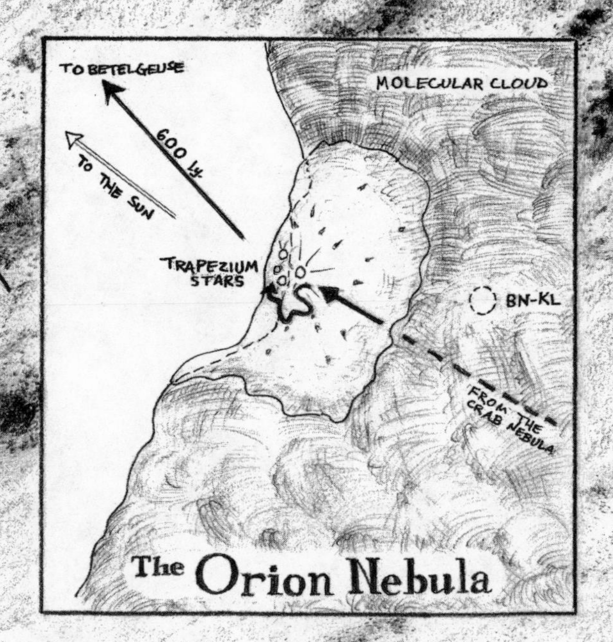

Prologue

If I have one complaint about the center of the Milky Way, it is that I found it nearly impossible to get a decent night's sleep. The sky teemed with blue-white stars, each as brilliant as a full moon. There must have been thousands of them, and I, sensitive to light as I am, felt pinned from all sides by their radiance. It made no difference that the portals of my craft could easily be covered to block out the glare; the knowledge that I was bathed in the light of this enormous star cluster was enough to keep me awake. Had I been able to sleep, I might have managed to convince myself that this long voyage was an entertaining dream, my answer to Johannes Kepler's *Somnium*. That great astronomer's allegorical excursion to the Moon, circa 1611, had been a favorite of mine as a student—I once translated it from the Latin myself—and since then I have often wondered what subliminal role it played in triggering my wanderlust. But this was no dream. The sensations were too acute and the discomfort too real to be so easily dismissed as imaginary.

It's funny how I dwell on the memories of such small annoyances nearly as often as I mull over the great themes that have emerged during my travels. In my mind I will always be able to revisit the searing, metallic glare that confronted me near the Crab Nebula's pulsar; the infuriating, impenetrable shroud of dust that blocked my explorations of the giant star Betelgeuse; the seasickness that overtook me as I surfed through heavy turbulence to the nucleus of the galaxy from which I now write. The grander episodes have been internalized, I suppose, and no longer serve to mark the specific events of this journey so much as to characterize my own progress as its witness. Among these episodes were the life-and-death encounters: a terrifying incident near the second black hole I visited, my escape from a supernova explosion in the Magellanic Clouds, the barrage of rogue boulders that nearly destroyed my craft as I tried to watch the creation of a planet. These dramas, any of which could have cost me dearly, I seem to have taken in stride. Each has helped to crystallize an aspect of this journey—to broaden my appreciation of relationships among phenomena that I scarcely imagined could be connected in so many ways.

Funny, too, that a few periods of enforced wakefulness should have disturbed me. After all, 85 percent of my journey has been spent in hibernation—a necessity, given the journey's length. I have been traveling, now, for more than 200 years by my clock. I am—or was—an astrophysicist of the early twenty-first century. I think back on that time frequently, and it is only with the greatest difficulty that I can accept the fact that (almost certainly) every last one of my colleagues has been dead for 60 million years! This trip has occupied less than half of my conscious lifetime, and part of my psyche cannot grasp the passage of so much time on Earth.

Yet there is nothing paradoxical about the distortions of time that I have had to create in order to make this trip. Because my craft has accelerated to within a hair's breadth of the speed of light (without ever exceeding it), small amounts of time pass for me while eons elapse for Earth and most other bodies in the cosmos. I am a manipulator of time, though I am doing nothing that is not readily comprehensible to every physics student on my planet.

Nor are the other technical aspects of my craft anything more than extensions of the technology that existed at the time of my departure. True, I pioneered the experiments that led to this highly efficient form of propulsion, and I was the first to build a prototype. But by the spaceship design standards of my time, my craft boasts few extravagances. No one would have called this a cathedral to space travel; it is certainly no space-metropolis.

Why should it be, given that I am its only inhabitant? My craft is cozy but not cramped. I have grown into it. I have lived here so long that I have come to think of it as an extension of myself, just as a hermit crab must come to regard its adopted shell. I'm organized but not especially tidy; fragments of my writings and calculations are strewn about. And I have set several view screens to project images that please me—reminders both of Earth and of places I have encountered on my odyssey. In a word, my craft is home.

This vessel has flawlessly provided the necessities of life for all my waking years aboard. Food and air have never posed a problem. I have modest stores, I recycle waste, and whatever else I need I synthesize from the material of space. The fuel scoops, which can be extended out to thousands of kilometers from my living quarters, sweep through a kilogram of interstellar matter per second of Earth time (far more per second of time as measured by my clocks). Most of this matter is hydrogen or helium, and it is instantly converted into energy for propulsion. A small fraction—about 1 percent—consists of oxygen, carbon, and all the other elements. Of this, only a tiny amount needs to be diverted for conversion into food or air or for the fabrication of any other materials I require (such as tools, paper, and patches for my craft's skin).

My craft is sturdy, too—I have survived, after all, and am in reasonable health—though it is perhaps not quite so sturdy as I should have liked. I have subjected it to conditions it was never meant to endure and have watched, horrified, as its outer layers began to boil and slough off under the intense pressures of an onrushing nebula. But the shielding has always held. Because it has protected me time after time, I have begun to develop the kind of affinity for it that comes with shared adversity—as though it were a traveling companion.

Does my craft have a name? I had never thought to name it—such a symbol of personification seemed unnecessary. But how about *Rocinante*? The name has some sentimental value for me. One of my graduate students gave this label to my research group's first computer. I immediately recalled John Steinbeck's motorhome in "Travels with Charley," but it turned out that she had named it after the spaceship in a rock group's popular song. My faculty colleagues assumed we had named it after Don Quixote's horse and proceeded to call the department's successive computers Dulcinea, Panza, and so on. Personally, I have always leaned toward the Steinbeck association, and I think it also suits my craft. *Rocinante* is homey in a rough sort of way, like a camper van. An old camper van, for that matter, with its fraying upholstery. In lieu of human companionship (always pragmatic, I have long since trained myself to cease regretting its absence), my craft at least provides the companionship of a stable, human-scaled environment that I have shaped. It is necessary armor against a cosmos that does not know I exist.

• • •

I see no reason to rehearse in too great detail the events that led up to this trip and have now receded so far into the past. I was motivated by several considerations. First, I traveled because I could. The craft was ready, and I grew impatient waiting for the formal cycle of testing and refinement that would have extended (I feared) far beyond my lifetime. Then there were academic and political disputes—over my experiments and my theories, over priorities and precedents—that now seem so petty and incomprehensible that they do not bear repeating.

The scientific and philosophical motivations remain much more vivid, although even they have evolved so far beyond recognition that it scarcely seems worthwhile to describe them exhaustively. A number of bizarre objects and phenomena had been the focus of my studies for several years. There were quasars, pulsars, jets, and black holes, to name a few of the more exotic ones. Were they all merely curiosities, or did their similarities, differences, and relationships convey some more subtle message about the way things are constructed and the way everything "works" out there? In what ways did they connect—if at all—with the "everyday" phenomena of the skies: the stars, galaxies, and nebulae with which I had long been familiar?

My scholarly interests in these phenomena, and the questions they raised, gradually took on the character of an obsession. Then, the invention of a means to travel in the realm of these bodies made the prospect of some sort of trip irresistible. To gather firsthand evidence about them—that became the objective against which I would have to judge the success of my work, eclipsing the accolades that began to rain down on me as word of my propulsion experiments leaked out.

I was enthralled with the idea of touching the places I studied, or coming as near to touching them as I could. For years I had endured the gloating of my colleagues in the field of planetary science. They had grown accustomed to visiting their objects of study—at least the local ones—and literally scooping up samples, if not in person, then through the agency of automated probes. We astrophysicists had always had to make do with remote viewing and indirect deductions based on highly idealized models for the phenomena we studied. A slight mistake in the observations, and we were easily fooled, as we were when we mistakenly attributed the pulsations of a nearby binary star to a quasar nearly coincident with it on the sky. To make any progress at all, we applied a "principle of mediocrity," asserting that any type of phenomenon found nearby should be characteristic of the Universe at large. This principle often went under the more dignified name of the Copernican Principle (after the Polish astronomer who had the audacity to lump Earth in with the other planets, removing it from its Ptolemaic position at the center of the universe). To extend the Copernican Principle to physical laws was to assume that those laws applied equally everywhere in the Universe. A very healthy conservatism, I suppose, that usually stood us in good stead, but a philosophy that made it much more difficult for an astrophysicist to discover a new law of nature.

Even worse off were the poor cosmologists, who had only one "object" to study. At least we could compare different galaxies and nebulae, much as the planetary folk practiced "comparative planetology." I often suspected that a kind of cosmic loneliness was the real reason why cosmologists were so apt to hypothesize parallel universes, "many worlds" theories, and the like. These alternative universes, patently undetectable, were like their imaginary friends.

The idea of bringing to these studies even the slightest bit of immediacy, of direct sensation, became my crusade. Finally I hatched a concrete plan: to travel to the center of the Milky Way, to see what was there, and then to return. Why visit the Galaxy's center? As best I can reconstruct it, I had fallen prey to a simplistic optimism about the root of all structures—a kind of cosmic Utopianism. I came to believe that the secret to the organization of the Universe could be summed up in a single word: gravity. Experience as pure a gravitational field as possible, and you will have had all the experience you need, I thought. At the time this seemed to make sense. Gravity was always attractive, unlike the forces that governed electricity and held sway inside atoms. And unlike these other kinds of forces, gravity did not falter (although it grew steadily weaker), no matter how large the distances involved. It held the Moon in orbit around the Earth, the Earth around the Sun, and the Sun around the Galaxy. The analogy went all the way up to the entire Universe, which actually could have recollapsed as a result of its own gravity, if only it had been sufficiently massive and had not flung itself apart so violently at the outset. Universal attraction seemed to be the key. Perhaps this is why, in a perverse and symbolic way, I welcomed the unpleasant sensation of being pinned inside the cockpit of my craft by the oppressive glare of the Galaxy's central star cluster. The stark shafts of light, beating in on me from all sides, immobilized me, as though I had fallen under the influence of some immense gravitational field. In a sense, this was what I had been looking for, although I had expected to find it manifested in a physical rather than a psychological force.

Is there a better place to experience pure gravity than in the vicinity of a black hole? After all, a black hole is often described as the disembodied gravitational field left behind when matter is sucked out of this universe and goes . . . who knows where? I knew that there was a good-sized black hole smack in the middle of the Milky Way. The evidence was incontrovertible. My colleagues—two groups of them, working independently in Germany and California—had plotted the motions of stars and found that their orbits were behaving as though they were under the gravitational influence of matter equivalent to 2 ½ million Suns. My jaw dropped when they showed me their "movie" of stars careening around in tightly prescribed orbits, half a Galaxy away. It had taken them decades to compile the frames, each one a year's worth of work. They looked in vain for a tight cluster of stars to provide the inferred attraction, but all they came up with was a faint speck glowing a bit with X-rays, radio waves, and gamma rays. If this wasn't a black hole, they concluded, it was something even weirder.

The Milky Way's center lay 26,000 light-years from Earth, but the trip, I calculated, would take just 20 years in each direction, by my clock. Because I would spend most of that time in hibernation, it did not seem that this journey would constitute such a large investment as measured against my life's duration. As measured against its texture, though, I knew the impact would be incalculable. I would return more than 50,000 years of Earth time in the future—sheer madness, from the point of view of resuming any sort of normal existence—yet my compulsion to travel was such that this does not seem to have worried me.

As it turned out, I did reach the center of the Milky Way, but then my plans went awry. Disappointed and confused by what I found there, I traveled onward, in search of some kind of closure to my mission, then onward again. Each leg of the journey drew me inexorably to the next and changed the character of my quest. A list of destinations and half a lifetime away from home: those are the facts of my journey. But such a list does not express the deeper structure that slowly asserted itself. That will take many more pages to describe.

Part One

Gravity

1

To the Center of the Milky Way?

No sooner had I decided to go than I began to get cold feet. This wasn't fear of *Rocinante*'s reliability—I was confident of my craft's propulsion and life support systems. It was not even fear of the unknown—in my arrogance, I thought I knew what I would find. This was fear of the unseen, the terror that strikes fogbound drivers: I could not see where I was going. To a casual viewer on Earth, the center of the Milky Way seems a nonexistent destination. If you don't believe me, go outside during July or August and try to find it for yourself in the night sky. It is easy enough to trace the Milky Way's luminous band as it sweeps through the ancient constellations of Cassiopeia and Cygnus, soars across the equator in the constellation Aquila (the Eagle), heads southward through the obscure star pattern of Scutum, and crosses the ecliptic—the path of the planets—in the zodiacal realm of Sagittarius. What appears as a band on the sky is really the cross section of a vast slab of stars seen from within, from a site near the midplane. People had figured out that much in the eighteenth century. Astronomers later determined that the slab is really a disk, complete with a far-off center. But the location of the latter is anything but obvious.

Scutum looked like a promising site for the kind of glitter one might expect to find at the pivot of a major spiral galaxy, so I decided to study its environs more closely. It certainly ranks as the constellation with the most arcane etymology. Scutum Sobieskii—Sobieski's Shield—is the only constellation named to honor a flesh-and-blood military hero, John Sobieski, who saved Poland from Swedish domination in the seventeenth century and then went on to defeat the Turks at Vienna. It is suitably decorated. With simple binoculars, I spotted the brilliant "open" star clusters Messier 11 and 26, looking like sprays of small diamonds scattered on black velvet. But these attractions proved to be scarcely outside Earth's neighborhood. Farther south, near Scutum's boundary with Sagittarius, I noticed two enormous glowing clouds of gas: the Trifid and Lagoon Nebulae, more distant versions of the Great Nebula of Orion that I was to visit later on. Yet even these are not the unique kinds of markers one would expect to find at the geometric center of a gigantic stellar merry-go-round like the Milky Way. I was getting discouraged, but it also turned out that I was getting warm.

I was staring straight in the direction of the Milky Way's center but didn't realize it. What I saw was a warren of bright ridges and dark lanes, broadening and seeming to become more complex and convoluted as I traced the Milky Way's path southward into Sagittarius. There was still no sign of the center, but it was there all the same, hidden by a screen of dust.

In my defense, even the best astronomers of the early twentieth century had had a tough time determining the Milky Way's center, to say nothing of the Galaxy's size and shape. They thought they had it in 1920, when a Dutch astronomer named Kapteyn mistakenly concluded that Earth occupied the place of honor. But "Kapteyn's Universe" turned out to be no more than the local patch of galaxy surrounding Earth. Kapteyn had estimated the distances of stars in different directions around the sky, assuming that the dimmer stars were farther, on average, in inverse proportion to the square root of their brightnesses. (Light bulbs grow dimmer with distance in this way, he reasoned, and so should stars.) Counting the numbers of stars at different distances, versus direction, he built up a three-dimensional model of the Galaxy. This was a more detailed version of an approach Herschel had tried a century earlier. But for this technique to work, two conditions would have to be met: The mixtures of star types (colors, sizes) would have to be similar at different distances, and the space between the stars would have to be transparent to starlight. The first condition is true enough, but the second is false. What neither Kapteyn nor anyone else knew at the time was that much of interstellar space is filled with a haze of tiny dust particles that progressively obscure stars at increasing distances, blocking their light entirely beyond distances of little more than 5000 light-years. As a result, Kapteyn severely overestimated the distances to the fainter stars, which are dimmed not so much by their remoteness as by a pall of interstellar smog. What Kapteyn thought was the entire Galactic disk was really only the patch within 5000 light-years, just 20 percent of the way to the Milky Way's true center. It is no surprise that the stars looked symmetrically arrayed about the planet Earth.

Who would have guessed that the Galactic disk resembles a smoke-filled room? The dark lanes and "coalsacks" that you can see with binoculars—regions seemingly devoid of stars—are merely places where the dust concentrations are especially high. The particles themselves are not that unlike the particles that make up cigarette smoke. They are about the same size—a fraction of a micron, or a few percent of a hair's breadth—and are somewhat similar in composition. Many of them consist of "soot," mostly graphite and an admixture of hydrocarbons, with some very tiny particles of "sand" (silicon-based minerals) mixed in. This dust is pollution from supernova explosions, lesser stellar explosions called novae, and the evaporating outer envelopes of giant stars, and it readily accumulates in stagnant pockets of space, where there are no winds to sweep the Galaxy clean.

Fortunately, the Milky Way's dusty disk is less than 5000 light-years thick, so if you can't see along it very far, at least you can see out of it. That is exactly why the Milky Way forms a distinct band on the sky, and it is why Kapteyn's predecessors, going back to Thomas Wright in 1750 (and later to include the philosopher Immanuel Kant), had guessed correctly that the stars of the Galaxy were confined to a slab. Only by looking "up" and out of the disk did astronomers spot the telltale—and indirect—evidence of where the true center lies. It was Harlow Shapley who guessed that the spray of "globular clusters"—the 100 or so disembodied "droplets" of concentrated Milky Way, each containing as many as a million stars in a tight spherical packet—formed an extensive halo framing the Galaxy's center like a bull's-eye. Earth was far from the center.

By this point, preparations for my voyage were far advanced. The millennium had turned a few years earlier, public (and government) interest in my experiments had been growing, and I realized that I might soon have to cede control of my research facilities to an increasingly nervous group of backers. Time was pressing, yet for an instant I hesitated. I knew from abstract arguments where the Milky Way's center was, but I was frustrated at not being able to see it. Did I really dare to embark on a journey toward an invisible destination? My enthusiasm began subtly to wane. Then, at the last minute, I was granted the preview I needed to cement my resolve—a spectacular image.

My good luck stemmed from the fact that light rays are not infinitely thin beams traveling along precisely straight lines but, rather, are slightly "fuzzy" and always a little bit spread out, no matter how tightly one attempts to focus them. This fuzziness, which increases with the wavelength of the light, means that light can bend very slightly around sharp edges (the phenomenon of diffraction) and cannot be blocked by any object that is much smaller than the wavelength. Interstellar dust grains have sizes comparable to or larger than the wavelengths of visible light, and hence they block these rays rather effectively, but they are smaller than the wavelengths of infrared rays. Thus infrared radiation from the Milky Way's center passes through the dust unimpeded. I recalled that my colleagues who had mapped the motions of stars in the Galaxy's center had exploited just this selective transparency. With the aid of an infrared telescope orbiting above Earth's atmosphere (the nearest place where one could get an unmuddied view), I peered toward that spot in Sagittarius that had been pinpointed so painstakingly, by triangulation off the globular clusters, many years earlier. I now saw the disk of the Milky Way clearly, converging to a thin band with distance. The Galaxy's bulge became clear, framing the Galactic nucleus. And zooming in, a compact star cluster was unmistakable, as was the glow from the warmed gas clouds that surround it. This distant view rekindled my eagerness. But it hardly prepared me for the real thing.

2

En Route

When I recall my journey between Earth and the Galactic Center, I picture the clouds. I see my vessel darting between and through dense pillars of dust and chemical residue that blot out all view of the stars outside. I recall a vague sense of foreboding and anxiety as I visualize my passage through the darkest clouds. It was a bit like flying through a thunderstorm, but (for the most part) there was no turbulence, which was eerie in itself. There were stars—they must have been a prominent part of the scene—but somehow they didn't make so great an impression as the dark, imposing clouds.

I was edgy. Faced with mounting pressure, I had left in a hurry and somewhat surreptitiously. The enormity of what I was undertaking hit me fully only after I was en route, and I brooded over everything I might have forgotten. Were all stores and life support systems fully charged and operational? Had my accelerators been set correctly? Had I said my goodbyes to those friends and colleagues I knew I would miss? At these speeds, communication with Earth would be cumbersome at best, and I knew I was unlikely to receive responses to the reports I sent back, half-heartedly, during the first few years of the mission. I also could not forget the rivalries and jealousies that had plagued the support teams. Would someone from Earth give chase? I worried, even though I knew this was utterly impossible. The risk of sabotage, or a simple mistake, was more real. On guard for a malfunction and too excited to risk missing any of the novel scenery, I hibernated only sporadically. The repeated cycling of dormancy and wakefulness took a toll on my nerves that compounded my more mundane cares.

The terrain seemed familiar as I departed the Sun's vicinity. The region near the Sun is fairly open and not too dusty, consisting largely of warm to hot gas—by which I mean gas at 100,000 to a million degrees Celsius. Such temperatures are high enough to sublimate most of the dust, and even the individual atoms of hydrogen (which account for 75 percent of the gas, by weight) are broken up ("ionized") by the heat into their constituent protons and electrons. The Sun's immediate surroundings are not typical, however. It seems that a supernova, or some very hot stars, existed in this region a few million years ago, leaving behind a hot bubble of gas. Several hundred light-years from the Sun, I plunged into a more typical mid-Galactic environment: a zone of scattered clouds. If you are puzzling over these meteorological metaphors, I should point out that there is no true vacuum between the stars. Gas fills every cubic centimeter of the Galaxy, coexisting with the dust in the smoggy regions with a gas-to-dust ratio of 100 to 1 by weight. Some of the clouds have temperatures of 10,000°C, "cool" enough for dust to survive, although most of the hydrogen is ionized even in these clouds. Then there are clouds at only a few hundred to a few thousand degrees, and these are cool enough for the protons and electrons to recombine into complete hydrogen atoms. I learned to spot the regions in which the gas consisted of individual hydrogen atoms because they emitted a peculiar radio glow—always at the same wavelength of 21 centimeters, according to my receiver. The origin of this glow reminded me of the complexity to be found in atoms of even the simplest element. It is produced as the lone spinning electron that orbits in each hydrogen nucleus repeatedly flips direction under the influence of unimaginably tiny magnetic fields created by the central proton.

Interstellar clouds are very different from earthly clouds, which are demarcated by changes in the state of suspension of water vapor in the atmosphere. The interstellar variety is characterized by abrupt temperature changes at the cloud boundaries. Nature abhors a temperature difference, and the thermodynamic imperative tries to drive everything toward uniformity. Heat flows from the hotter surroundings into cooler clouds, causing them to evaporate. The breezes that blow everywhere in the Galaxy are perpetually tearing wisps off them. I wondered, then, why all this structure persists. If these clouds are continuously being destroyed, mustn't new ones replace them? Only later did I learn that the clouds are indeed a telltale sign of the great Galactic cycle of birth and death, made from gas squeezed in the shock waves of stellar explosions and from the dense envelopes of gas shed by slowly dying stars.

I gradually began to encounter some very chilly clouds, so cold and dense that the hydrogen atoms paired off into molecules. The 21-centimeter radio glow faded, quenched by the interatomic pairing, but it was replaced by a much more complicated spectrum of radio waves. These molecular clouds were truly imposing, some of them more than 100 light-years across and containing as much matter as 100,000 Suns. Some were bursting with clusters of bright young stars, but most churned quietly, with just a few modest-sized stars forming here and there. As I traversed these clouds, I encountered dust storms—even occasional "snow" (ice-coated dust grains)—in some of the denser pockets, where temperatures dropped as low as 10 degrees above absolute zero. In the rainbow of radio waves I detected the signature of toxic carbon monoxide, as well as other, more complex molecules: alcohols, formaldehyde, even water.

Halfway from Earth to the Milky Way's center—13,000 light-years from home—the concentration of molecular clouds noticeably thickened. Yet I still passed through an occasional clearing, nearly devoid of gas and dust where temperatures soared to a million degrees or more. These holes had been blasted in the interstellar cloud-deck by the combined effects of supernova explosions and winds from earlier generations of hot, massive stars. These clearings are so much more dramatic than the small bubble in which the Sun is immersed that they had been noticed from Earth and christened "superbubbles." As I looked "up"—out of the Milky Way's plane—while crossing these regions, I could see the faint X-ray glow where plumes of superheated gas forced their way out through the Galaxy's atmosphere.

But by two-thirds of the way to the center I was socked in, amid a nearly continuous layer of molecular clouds that was itself sandwiched between layers of cool atomic gas. Were these gaseous lanes part of the famed spiral arms? You may have heard of them. Disk galaxies such as the Milky Way seldom look like featureless platters when viewed from above. Often they sport grand spiral patterns that sometimes can be traced across the entire face of the disk, outlined by chains of dense clouds and dazzling clumps of newly formed stars.

What are these spiral arms? Certainly they are not rigid streamers of stars and gas, twirling around the pivot of the galaxy like a pinwheel. Astronomers had once speculated that they were organized structures held together, elastically, by magnetic lines of force, but it quickly became clear that gravity was the only agent capable of organizing matter over the large scales of a galactic disk. The main effect of gravity on a galaxy's disk is straightforward enough. Just as in the Solar System, where the nearly circular motions of planets balance the gravitational pull of the Sun, in a galactic disk the pull of gravity is resisted by the nearly circular motions of the stars and gas clouds about the galaxy's center. The planets farther from the Sun take longer to go around once than the planets closer in, and the same is true of stars and gas clouds in a galaxy. (The Sun takes about a quarter of a billion years to orbit the Milky Way once; the orbital time for other stars varies roughly in proportion to their distance from the center.) The gravitational forces so dominate other effects that you could no sooner maintain a rigid spiral pattern of stars and gas in a galaxy than you could keep the planets rigidly lined up in the Solar System.

There is, however, one complication in a galactic disk that does not apply to the Solar System. Unlike the Solar System, where 99 percent of the mass is concentrated in the Sun and the individual planets have minimal effect on one another's motions, a galaxy's gravitational field consists of the combined gravity of all its constituent stars and gas clouds. This can make for very complicated orbital motions, because the amount of matter a body is orbiting depends on how far it is from the center, on whether the matter is symmetrically placed about the center, and so forth. Moreover, it is not just the mass of the disk that holds the galaxy together. Theorists have shown what would have happened to the Milky Way if it had consisted of a disk alone. It would have been violently unstable, sloshing from side to side, even buckling, and ultimately losing its disk-like shape altogether. The reason why this does not happen is that the entire disk is embedded in a vast spherical halo consisting of an even larger number of stars; very tenuous, hot gas; and a third, catch-all component that is so hard to detect that astronomers still refer to it as "dark matter." In fact, most of the gravitational pull that keeps the Solar System from shooting off into intergalactic space comes from the halo, not the disk. The stars in the halo are mostly old and faint, which makes even the "non-dark" part rather hard to detect. But individual halo stars do occasionally give themselves away because they do not participate in the orderly orbital motion of the disk stars. As I traveled I found interlopers easy to spot, because their motions seemed to be completely random, and they were usually moving much faster than any of the disk stars.

Even with the halo in place, the galaxy's disk is still slightly unstable, and it is this slight tendency toward imbalance that generates the graceful spiral patterns. As in most great art, the slight imperfections are what count. I had always admired pictures of well-formed spiral galaxies for their aesthetic qualities, but I knew that the mechanism that leads to spiral structure is prosaic and explicable entirely in terms of the subtleties of gravity. Suppose that by some chance fluctuation, a group of gas clouds and stars bunch up a bit as they orbit the galactic center. The bunched-up matter exerts a little extra gravitational pull on the surrounding stars and gas clouds and slows *them* down slightly. They then bunch up, while the original bunched-up stars and gas clouds gradually return to their normal, clockwork paths around the galaxy. The newly bunched stars and gas then cause the next group of stars to bunch up, and so forth. A bottleneck is created and perpetuated: a classic rubbernecking delay.

Thus the spiral "arms" are more aptly called waves, because they are not really objects but rather patterns. They are merely the spiral-shaped loci where the stars and gas of the disk slow down briefly as they march around the galaxy's center—a kind of interstellar traffic jam. Traffic jams may one day seem quaint anachronisms on Earth, once vehicles are controlled by computer, but it is doubtful that traffic engineers will ever learn to tame gravity.

What make spiral arms so obvious are the "tracers" that outline them: chains of molecular clouds and clusters of newly formed stars. If spiral arms are traffic jams, then the tracers are the result of accidents caused by the backup. As the gas clouds bunch up, some of them run into one another and merge, creating molecular agglomerations larger and denser than average. With increasing size and density comes an increase in the local gravitational force. These giant molecular clouds are set with a gravitational hair-trigger, primed so that their own self-attraction will overwhelm them given the slightest provocation. Any sudden compression as two clouds collide will cause them to collapse under their own weight. Perhaps that's all it takes to trigger star formation: cloud collision and—voilà—spiral arms are bejeweled with brilliant strings of young stars.

As I mused on the nature of the thick cloud layers that surrounded me, it became clear that I had already left the spiral realm behind. These gaseous strata were too thick, too permanent-looking to represent anything as ephemeral and delicate as the passage of a spiral wave. I realized that I was already too close to the Galaxy's center to encounter spiral structure, and I recalled a map I had once seen, showing the positions of the Milky Way's spiral as deduced by radio astronomers. According to this map, if you were to look at the Galaxy from above the disk, you would see the arms splayed out openly in the outer regions of the Galaxy. Indeed, I would have crossed better-defined arms had I headed in the direction of Cygnus or Perseus, instead of toward Sagittarius. Closer in to the center of the Galaxy the spiral pattern becomes more tightly wound, less spiral-like and more like a series of concentric rings. I figured that I must have been passing through these rings, just about at the place where they begin to merge into an undifferentiated continuum.

Suddenly I understood that even this detail was a subtle part of the scheme by which gravity organizes the Galaxy. Gas passing though a spiral arm tends to lose just a little bit of its orbital motion because of the retarding effect of gravity, and consequently, it drifts closer to the center. Thus spiral waves help gravity to achieve its universal goal of attracting all kinds of matter—stars, gas, whatever—toward a common center. In these dense molecular and atomic gas are cloud-layers a few thousand light-years from the Galaxy's center, much of this drifting gas seems to accumulate.

I was now expecting a gloomy ride the rest of the way in. Gravity indeed would like to collect ever-denser banks of cloud and dust toward its central focus. Surprisingly, though, the gas doesn't drift all the way to the center. As I emerged from the innermost molecular band—a couple of thousand light-years from ground zero—the atmosphere was still murky, but I noticed that the clouds were becoming much patchier than in the dense smogbanks I had just traversed. A view of the scene in infrared rays showed why. I was now passing through the "bulge" of the Milky Way Galaxy, where the disk appears to puff up and meld with a sort of inner halo. An ever-increasing fraction of the stars around me were moving randomly, rather than marching in lockstep circular orbits. The Galaxy's geometric center lay dead ahead, but there was something strange about the pattern of stars on the sky. I hesitated for a moment, disoriented. But then I saw what the problem was. Instead of being symmetrically arrayed under the evenhanded attraction of gravity, as I had intuitively expected, the distribution of stars was lopsided! There were more stars to the left of the Galactic Center than to the right.

I was face to face with the Galaxy's stellar "bar," which resembles nothing so much as a huge tumbling peanut made up entirely of stars. Stars caught up in the gravity of the bar execute motions that are far different from the orderly circular orbits of disk stars, different even from the chaotic dashing of halo stars. These trapped stars trace out semi-repetitive shapes ranging from figure eights, to complicated cat's cradles, to woven tubelike structures reminiscent of those "Chinese puzzles" that trap your fingers. It is amazing that these patterns can hang together, but bars are remarkably robust and are found in a good half of the spiral galaxies.

Thanks to the bar, for once the inexorable central pull of gravity is foiled. The bar hinders the inward drift of gaseous debris. The churning gravitational forces produced by the tumbling peanut stir up the motions of any gas clouds that venture inside, driving the clouds slightly farther from the Galactic Center. That explains why I was emerging into a region of the Galaxy where the clouds were becoming sparser. But it does not explain what I saw next.

3

A Ballet

Suddenly I emerged into a clearing only 10 or 20 light-years from the Galactic Center and saw a cluster of stars the likes of which I had never seen before. Nearly a million stars were crammed into a volume that would have been occupied by only a few dozen stars had it been in the vicinity of the Sun. I was immediately struck by—and would lose several night's sleep to—the many blue-white, blindingly luminous stars that were mixed in with the more common stellar varieties in every direction. These kinds of superstars, though only a few times more massive than the Sun, were so bright that they were doomed to burn themselves out in a blaze of glory. I had seen a few of them en route from the solar neighborhood, but everywhere else in the Galaxy they seemed to be quite rare. This relative rarity was not a surprise, because such stars last less than 10 million years (compared to 10 billion years for a star like the Sun) before throwing off their envelopes in violent convulsions and, if they are heavy enough—more than 8 or 10 times the mass of the Sun—exploding as supernovae. But even before they become violently unstable, they will have spent most of their lives injecting fast, hot streams of gas into their surroundings.

Given how many massive stars there were, it came as no surprise that the whole Galactic Center was immersed in a kind of hot bubble, similar to the superbubbles I had traversed earlier but lacking their opportunity to vent into the Galaxy's halo. As a result, the pressure outside my craft had increased enormously, although it was still much lower than any vacuum ever produced on Earth. I had measured it when I arrived in interstellar space near the Sun: There it was 0.00000000000000001 of the mean atmospheric pressure at sea level on Earth. But in the Galactic Center it was thousands of times higher than in the Sun's vicinity, pumped up by the combined effect of the fast winds and the added jolt of a stellar explosion every thousand years or so.

How did these stars get here? Because they were burning up inside at such a prodigious rate, they couldn't have been older than a few million years. Were they formed here? I looked around for likely sites of star formation, giant clouds rich in molecules, shielded from the heating and evaporative influences of hot stars and supernova blasts. But I couldn't find any. Streamers of compressed gas stretched here and there, squeezed by the high pressure and combed out along magnetic lines of force. A few of the nearby gaseous streamers seemed dense enough to collapse under their own weights, and it's just possible that I spotted a couple of stars forming. But the impression I had was that the cluster occupied a vast, nearly empty cavity, the space between the stars glowing faintly in X-rays because of the high temperature, which now topped 10 million degrees in places. What gas clouds I saw were being flung about chaotically in all directions—not exactly the optimal conditions for organized gravitational contractions to create, or replenish, a cluster of nearly a million stars. Were conditions very different a million, 10 million years ago? Was the Galactic Center then full of dense, dusty clouds of gas, churning out all these stars in a giant burst of gravitational coagulation?

I still do not know the answer to this question. The hypothesis offended my Copernican roots, because it would have meant that I was visiting the Galactic Center under special conditions, and it raised more questions than it answered. Why should I have happened upon the Milky Way's nucleus at just the "moment" (astronomically speaking) when rampant star formation had stopped but its shortest-lived products were still vital? Why was there no molecular gas in reserve anywhere near the present-day star cluster? I searched desperately for alternatives. Then, with little warning, one was thrust upon me: At least some of these young, massive stars must have formed from the collisions and mergers of smaller stars.

Watching two stars collide and coalesce has to be one of the most spectacular sights in the Galaxy, but you have to be lucky to see it. Stellar collisions occur only once every few thousand years, and only in the centers of galaxies are the stars packed tightly enough for collisions to occur even this frequently. It is a safe bet that no stellar collision has ever occurred in the environs of the Sun.

The whole concept of a stellar collision sounds violent, but what I witnessed was akin to a ballet. The dance begins tentatively, for only in rare cases do the stars hit head-on. Most often they barely brush one another and, if conditions are just right, glide into a delicate embrace. The collision, then, starts as a kind of *pas de deux*. The stars in question approached one another more slowly than I expected, at not more than a couple of hundred kilometers per second. I should not have been too surprised, because such speeds are typical of the stars located a few light-years from the Galaxy's center of mass. But slowness was a crucial element in the encounter. If the mutual approach had been too fast, these stars would have sped past one another with little interaction, or (if aimed just so) they would have hit so hard that the outcome would have been ugly: bits of star splattered everywhere, but no long-term relationship.

As the stars glided toward one another, their motions were gradually deflected from straight-line indifference to gently converging paths, and they sped up. I noticed the swellings that rose gradually on the sides of the stars facing one another, as well as on the opposite sides. Through these bulges, each star was responding to the other's gravity, which is stronger on the near side than at the star's center, and stronger at the center than on the far side. (On Earth we get excited about the barely perceptible tides caused by the Sun and Moon, but they pale beside these distortions.) By now the partners were focused squarely on one another, stretched along their mutual axis. As they swung past each other, the bulges tried to follow, and for the most part they kept up. But stars do not like being distorted, and the deformation took its toll. The friction of each star's continuous internal readjustment heated its interior, and the bulges began to lag behind the stellar motion. The bulges no longer lined up, and their gravitational attractions for one another slowed the stars and drew them closer together. I held my breath, because I knew that this was the crucial stage in the encounter. More often than not, capture would elude the partners: They would release their gentle grasps and swing apart. But in this case the bulges lagged far enough behind, and gravity had enough leverage: The stars just managed to swing into orbit around each other.

Like most stars that have newly captured one another, my couple shared a graceful orbit, swinging far apart and then plunging close. When far apart, the stars seemed almost oblivious to one another—in this phase, indeed, passing stars have been known to steal other stars' partners. But when they plunged close together, the bulges reappeared, and the stars sank deeper into one another's gravitational influence. By now the incessant heating had inflated the stars' atmospheres, and their gaseous envelopes had begun to mingle. The stellar nuclei, where nuclear reactions pump out energy, were still distinct, but they were rapidly being subsumed beneath the common envelope. Finally, they merged.

According to theory, it would take another 1000 years or more for the merged star to settle down. The "new" star would become much brighter than the sum of its two progenitors, and not just because the merging process itself generates a lot of heat. The nuclear reactions that power stars are *extremely* sensitive to temperature (hence the term *thermo*nuclear), and the temperature inside a star depends on the star's mass and size in a way that is determined by the necessity of a balance between gravity and pressure. Thus, if the new star is double the mass of the old, it must be roughly twice as large. If it were too small, the temperature in the center would be so high that nuclear reactions would proceed at an explosive pace, the pressure would build up, and the star would expand. If it were too large, the central temperature would be so low that the nuclear reactions would fizzle out, and the star would contract. In this way, the temperature sensitivity of thermonuclear reactions provides an elegant feedback that determines the sizes of stars. But a different effect determines how bright a star is. Energy leaks out faster from a larger star than from a smaller star, because the former has more surface area and is generally more porous. As a result, more massive (and therefore larger) stars put out *a lot* more light than low-mass stars. A star only twice the mass of the Sun puts out about 16 times more light. The flip side is that a 2-solar-mass star has only twice the fuel supply of the Sun, so it can live only 1/8 as long. When I put the arguments together this way, it no longer seemed like such a crazy idea that the hot young stars in the Galactic Center Star Cluster might well have formed from mergers.

My theoretical training allowed me to anticipate the future of this newly merged star, but I had no time to watch the final stages unfold. I hadn't come here to study stars, anyway. I was searching for pure gravity, and pure gravity was to be found in one place: the big black hole at the very center of the Milky Way. But how was I to locate the Galaxy's exact center of mass? I looked about for some secondary clues and noticed, in one direction, an especially dense concentration of stars surrounding a point of light with a strange, very blue glow. As I headed toward it, I immediately noticed that the stars around me were getting closer together. Their random motions were also speeding up: 500 hundred kilometers per second, 1000, 1500. . . . Any stellar collisions that occurred here would be far from gentle. They certainly would not lead to graceful mergers. The stars would be smashed, their debris dispersed to interstellar space. Could that be where some of the streamers of gas had come from?

I pulled out my calculator and started taking notes on how the stellar speeds were increasing as I approached the blue glow. At 1/10 of a light-year the speed was 600 kilometers per second, at 1/20 of a light-year it was 850, and so forth. Every decrease in distance by a factor of 4 brought a doubling of stellar velocity. I quickly deduced that gravity was increasing just as Isaac Newton had predicted for a body with a mass two-and-a-half million times that of the Sun, all concentrated in one place. I was clearly sensing the black hole's gravity. Yet my initial sense of satisfaction faded as I recalled how my colleagues had deduced everything I was finding, from the mass of the black hole to the shape of the star cluster, without leaving the comfort of their observatory 26,000 light-years away. (Funny how one never focuses on one's advantages in these situations. For example, it never occurred to me to gloat that my colleagues hadn't witnessed a stellar merger.) In any case, I had no time to wallow in these conflicting emotions. If I had not braked hard and gone into orbit about the Milky Way's central black hole, I soon would have become part of it!

4

Ground Zero

I was surprised that my arrival close to the Milky Way's central black hole did not strike me more viscerally. There was never a point at which I could feel, between my head and feet, the stretching force due to the stronger gravity closer to the hole. I knew that these bigger black holes are altogether more gentle on visitors than their smaller counterparts, yet I expected some gut feeling to tell me that I was in the presence of an enormous source of attraction. The visuals were hardly as dramatic as I expected, either. There was no grand disk of superheated gas, crackling with energy, swirling into the black hole. All I could see was a diaphanous, bluish luminescence surrounding what I deduced to be the black hole's location. I could tell where the hole was by observing the distortions of the stars beyond it as their light rays curved while crossing the hole's gravitational field. A lens-like distortion of a distant stellar scene, some subtle gradations of the blue glow—was that all there was to see of the Galactic central black hole?

Yes, I was disappointed, but I should have seen it coming. What made this black hole appear so serene was that it resided in a near vacuum. There was no "donor" star to be plundered for its substance, as I later encountered while visiting the much smaller black hole known as Cygnus X-1. The dense interstellar clouds were few and far between, and they were so stirred up by all the hot stellar winds and supernovae that they responded little to the black hole's lure. Quite simply, there was very little matter flowing into the black hole at the center of the Milky Way.

I ventured into the outer reaches of the blue-glowing corona. The gas there was so tenuous that I felt safe enough immersing my craft in it. I measured the radiation and found that it was not really "blue." It merely appeared blue when filtered through my visual range. Electromagnetic radiation was being produced across the entire spectrum, from radio waves through microwaves, infrared, visible, ultraviolet, X-rays, and gamma rays, the bulk of it coming out in the infrared and ultraviolet. I tested the temperature of this gas, which was a measure of the energies (and thus the chaotic motions) of its constituent particles. I was still hoping to accumulate more evidence of the effects of gravity pulling the gas inward, thereby causing the particles to move faster. As expected, the temperature seemed to be going up about inversely with distance from the black hole; it reached a billion degrees when I was roughly as far from the black hole as Pluto is from the Sun. But my instruments did not seem to be operating quite as reliably as usual, and I began to record unsettling temperature swings.

The reason proved to be simple, though bizarre: This gas did not have a temperature, in the sense in which temperature is usually understood. At such low densities and high speeds, it is difficult for particles to share their energies with one another. The roughly equal sharing of energy among all particles in a gas is the hallmark of "temperature." In Earth's atmosphere, this kind of sharing occurs so instantaneously that one can be sure an average oxygen molecule will carry the same amount of energy as the average nitrogen molecule and will therefore be moving with only 93 percent of the nitrogen molecule's speed (the square root of $^7/_8$, which is the inverse ratio of their molecular weights). But here, I found that the electrons had wildly different—usually lower—energies from the protons, despite the fact that all the particle varieties were mixed together. To make matters even more confusing, I found that a small fraction of the electrons had vastly higher energies than the protons.

This weird distribution of energies went hand in hand with the weird spectrum of radiation. I pictured the glowing heat-shield of my craft on the innumerable occasions when I had re-entered the Earth's atmosphere in early tests. That progression of colors—red, then orange, yellow, and blue-white—is burned into the visual memory of any habitual space traveler, as is its significance: It tells you the heat-shield is getting hotter. Any solid substance glows with a fixed shade of color that depends on its temperature. The same goes for dense gases, like the atmospheres of these blue stars in the Milky Way's center (20,000–30,000 degrees), the atmosphere of the Sun (5800°C), and even the Earth's atmosphere (glowing in the infrared at about 300°C above absolute zero). But the gas surrounding the Milky Way's central black hole is so transparent, so tenuous, and so "non-thermal" that its "color" seems to have nothing to do with its temperature. Or maybe it is more precise to say that it has no well-defined color. Or, if you will, it is so blessed with electromagnetic radiation of all kinds that it can't decide which color to be.

I followed this multicolored whiff of glowing gas as far I could toward its doom. In the interest of safety, I took care not to venture below the 24-million-kilometer "orbit of instability." This is not the horizon of the black hole, below which there is no way to escape—that's at 8 million kilometers. But according to Einstein's general theory of relativity (the theory that describes black holes), it is as close as one can orbit without constantly firing thrusters to keep from falling in. I had no confidence that my piloting skills could keep me out of trouble if I went closer. The corona at this distance exhibited less orderly motion than I was later to find in the disk of Cygnus X-1. Gas was rushing around in all directions. In some sectors it was plunging inward; in others it seemed to be blasting outward in a comparably chaotic rush. Superimposed on these turbulent eddies, the gas was swirling around the hole faster and faster the closer I got. The pressure was so high that I expected it to enhance the effect of gravity in sweeping gas into the hole at a prodigious rate, by pushing inward as gravity pulled. Yet just the opposite seemed to be happening. As the gas was pressed toward the hole, it only swirled faster, and that just seemed to make it stiffen, somewhat like hard rubber. The stiffness of the gas seemed to be holding it back, enabling it to resist the black hole's lure even as it crossed the 24-million-kilometer threshold from which I watched. Where did it begin its final plunge? I peered toward the hole and barely made out, at about 16 million kilometers from the hole, what looked like a sudden drop in the glow from the gas. I convinced myself that this was where the pressurized resistance to gravity failed and the gas thinned out as it was finally sucked in. But let me be honest: The appearance was so subtle and nondescript that to this day I do not know whether I saw the Milky Way's giant black hole swallow anything of substance.

There is an old saying among astronomers that black holes are the Universe's most efficient engines. But what I found in the Milky Way's center seemed to give the lie to this old dictum. What the saying meant was that you could get the most energy out (in whatever form: light, heat, jet propulsion. . .), for the least amount of fuel, if you let the fuel spiral into a black hole. Of course, there was no way that this energy could come from *inside* the black hole; the idea was that the very hot gas would release its energy just before sinking beneath the horizon.

It may seem odd to worry about the fuel efficiency of a black hole when the fuel requirements of any astronomical object are . . . well, astronomical. To power the Sun with nuclear reactions, for example, 620 million metric tons of hydrogen must be converted into helium every second in the Sun's interior. Yet if the Sun were a giant coal furnace, one would have to shovel in more than 20 *thousand trillion* tons of coal per second to achieve the same power output. Thus nuclear fusion is more than 30 million times more efficient than coal power. Instead of looking forward to a life expectancy of 10 billion years (of which 5 billion have passed already), during which it will use up only 10 percent of its supply of hydrogen, the coal-powered Sun would incinerate itself completely in 3000 years. Common wisdom had been that "only" 43 million tons per second of any kind of matter would need to be fed into a black hole to produce the Sun's luminosity, and this would make gravitational power 15 times more efficient than its nuclear counterpart. Indeed, I later found a black hole in which energy was being released with such high efficiency.

The big black hole in the center of the Milky Way was not in this league. Granted, the gas drifting into it was releasing 1000 times more energy per second than the Sun, but compared to the black hole's mass this was a pittance: The hole weighed more than a million suns. And it was gobbling up not 43 billion tons of fuel per second, as a "fuel-efficient" model would have, but thousands of times more. Surely all that matter was releasing an enormous amount of energy as the black hole's gravity compressed and accelerated it. But only a tiny fraction of that energy was getting "out." Where was the other 99.99 percent going? As I pondered this, I realized why I had found it so difficult to spot the demarcation between the gas swirling through the corona and that rushing in toward the black hole. Instead of shining brightly, the former was nearly invisible because it was jealously guarding its store of energy. Much of this energy was being dragged into the black hole, never to emerge. And some—perhaps most—was being shot back out into space as a superhot wind, helping the massive stars to stir up the Galaxy's central cavity.

I was discouraged. I had set out to learn all about the organization of the Universe by studying one big black hole, but the one I had picked to study had turned out to be a dud. What was clearly "wrong" with the Galactic Center black hole was that it occupied an anemic environment. For whatever reason (the powerful winds from all those hot stars, the gravitational deflections of the bar—it didn't really matter what the cause was), the effect was that this black hole was not eating enough to make it interesting. Sure, I could measure the increasing force as I approached the hole; I could see how it sped up and concentrated the motions of stars; I could even study the distortions of rays of starlight as they, too, felt the effects of the hole's gravity. And I did see something of the enormous capability of a black hole to set gas in motion and got an inkling of its ineffable ability to endow inbound gas with so much energy that the most violent thermonuclear reactions pale in comparison. But I realized—with a touch of irony—that I had learned more about gravity from other structures I had seen than from the black hole. The odd "tumbling peanut" of the Galaxy's bar and my musings about spiral arms had shown me the crucial role of motion in opposing gravity's attractions, not to mention the subtle interplay by which gravity works with motion to create structures. And despite its relatively puny energetics, the collision and merger of two ordinary stars had put on a much more spectacular pyrotechnical display than anything I had seen the black hole do.

I was desperate to find a way to salvage my mission. I began to experience bizarre daydreams, in which I visualized various ways to make the Milky Way's black hole light up. I imagined all kinds of strange schemes to bring a large quantity of gas into the immediate vicinity of the black hole and to dump it in all at once. Subjected to rational scrutiny, most of these ideas betrayed the kinds of inconsistencies that daydreams usually suffer. But some seemed half plausible, perhaps informed by scholarly tracts I had read before my departure. One in particular seemed highly realistic. What if the black hole swallowed an entire star?

My daydream began with a vision of those stretching forces I had searched for in vain when I first descended toward the black hole. These had proved too weak to be discernible by a person, but the same cannot be said for ordinary stars that venture too close to the hole. Even though it weighs as much as 2.5 million Suns, the Milky Way's central black hole is too small, too concentrated to swallow an ordinary star whole. Instead, its gravitational field will pull apart any star that comes closer than about 10 times the radius of its horizon, or about 80 million kilometers.

As I drifted off, I saw in my mind's eye a doomed star, approaching from out of nowhere. Of course I knew it was really on a highly elongated orbit, jostled into making the plunge by its gravitational interactions with other stars. As it approached, I observed the familiar tidal swelling rise on the star, an effect due this time to the difference in the black hole's pull on the star's near and far sides. But in contrast to the stellar merger, where the tidal forces had had a gentle touch, here the distortion got greater and greater the closer the star came to the hole. Eventually, the star's gravity could no longer hold it together, and it simply came apart. But not in a gentle way: The black hole's gravity ripped it to shreds.

The dance metaphor, perhaps overworked when I tried to describe stellar mergers, popped back into my head. I saw the debris of the star performing an elaborate dance, only this time it was more like a tango than a ballet. About half of the material was flung off into space at high speed, a shimmering spray, never to return. The rest returned to an elongated orbit, but now as a curtain of matter that swung far from the black hole, then fell back into its grip. I wondered whether it would ever settle down; for a time, it swooped in and out, on ever more surprising trajectories. But settle it did, and what remained in the end was a thick donut of gas, swirling in a regular fashion. It was rotating too fast to fall into the hole, but as I watched, it gradually spread inward as well as outward. Eventually, the inward side reached the black hole and erupted in a blaze of glory. This time, there was no question of all the energy disappearing into the hole. Briefly, the Milky Way's central black hole shone, in my imagination, with a power a billion times greater than its normal blue glow.

My alertness returned; I felt refreshed. Should I wait for a tidal disruption to occur, as it must eventually? No plunging stars were in sight, and I certainly could not wait the 100,000 years—maybe more—until a star committed suicide in this spectacular way. I decided to leave.

I now realized that it wasn't sufficient merely to witness gravity's potential to attract and to set matter in motion. One had to pay careful attention to the nature of that motion if one really wanted to understand how things worked. I should have seen the hints earlier. Every time I had focused on the organizing power of gravity, I had been drawn to its influence on the motions of things, whether it be the eggbeater effect of the Galaxy's bar, the self-sustaining sweep of the spiral arms, or the violent encounter of star with star or star with black hole. No matter how important gravity was, ultimately, as the root of cosmic structure, just floating in space and contemplating a huge black hole was not going to teach me everything I needed to know. I had to find a more dynamic system to study.

Part Two

Motion

5

The Cannibal

Legends had circulated for years about the existence of black holes that constantly gorged themselves, in a kind of cannibalistic ritual, on nearby dying stars. They were said to draw in matter so ravenously that they positively sparkled with X-rays and flickered at rates that could have stroboscopically frozen a hummingbird's wings (if one could have illuminated a hummingbird's wings with X-rays). The strange thing—and here is what made it so hard to believe—was that these black holes, unlike the one I had just visited, were not massive sinkholes lording it over an entire galaxy These well-fed black holes were reputed to be the mere remnants of ordinary stars, barely a few times more massive than the Sun. And yet the pathetic, gaunt aspect of the starved black hole at the Milky Way's center haunted me. I had just seen a million-solar-mass black hole, in the center of it all, not even commanding enough nibbles to keep itself shining brightly. How could these lightweights possibly maintain their gluttonous habits? Deeply skeptical, I set out to investigate.

Being a conventional tourist at heart, I set my course for the first black hole that comes to anyone's mind: the granddaddy of them all, Cygnus X-1. This system is located about the same distance from the Milky Way's center as Earth is, but in a completely different direction; it is roughly 6000 light-years to the east of the Solar System, as measured along the midplane of the Galaxy's disk. As I drifted into hibernation, I clearly recalled the heady days of the 1970s, when we had pored over each new piece of data as it came in, trying out those convoluted, indirect arguments. Was it a black hole, or not? Yes, it was so luminous that it had to be at least as massive as the Sun; otherwise, its own radiation would have inflated it so grotesquely that it would have burst (or suffered some similarly disgusting fate—we weren't sure what). Yes, it was emitting X-rays and was flickering so fast that it had to be very small. Less than 300 kilometers across. Remarkable. Ah, but neutron stars had been discovered by then, and they were *almost* as compact as black holes. Could it, just possibly, be a neutron star? And then the clincher. That 5.6-day orbital period, the discovery of the companion star: It was a member of a binary system. Then, even more remarkably, the measurement of how fast the companion was moving, or at least how fast it was moving toward or away from us (any sideways motion, along the sky's dome, could not be detected), which enabled one to say something about how heavy the X-ray emitter was. The "mass function" of Cygnus X-1 became the talk of every astronomy department canteen. It was really a lower limit to the mass of the "compact object," and it did rely on one's having correctly interpreted the companion star's speed, but it was a big lower limit—more than 3 solar masses.

What had been seen, when boiled down to essentials, was encoded in the object's name. Cygnus X-1 was the premier source of X-rays in the entire constellation of Cygnus, the Swan that glides along the Milky Way on a summer's night. It was tiny and too massive to be a neutron star, the only other kind of object that could approach it in compactness. By the process of elimination, it had to be a black hole. That was as well as one could do in those days, I thought, as I approached the system for my first close-up look.

I knew, of course, that the binary nature of Cygnus X-1 did more than provide a convenient method of estimating its mass.

By this point I understood that it was the interaction of a black hole with its surroundings that was the key to anything observable. I had learned that lesson painfully as I retreated, in disappointment, from my encounter with the sterile Galactic Center black hole. Cygnus X-1 shone so brightly *only* because it was devouring its companion. But how could two stars, presumably born in binary harmony—maybe even twins—have descended into such an atavistic and violent relationship?

What was clear was that the two partners of the Cygnus X-1 system must have completed an elaborate dance long before I arrived. I recalled the numerous articles that had speculated on this distressing topic. It was very likely, indeed, that the two members of the Cygnus X-1 system had been born together. They were fraternal, not identical, twins, for it was the remnant of the originally heavier twin that was now dining on its sibling. What was now the black hole had once been an ordinary star—30 or 40 times more massive than the Sun, hotter and a million times more luminous, but otherwise similar. Before its collapse to a black hole, it had spiraled so close to its companion—at the same time swelling in the normal decrepitude of stellar old age—that the two stars had mingled, with the dangerous result that most of the matter in the proto-black-hole had drained into its partner. Now that the matter was draining the other way, I wondered: Was this the payback for earlier sibling treachery?

It is usually assumed that the more massive star of a coeval duo will meet its maker first, because more massive stars burn up more quickly. Therefore, you will be forgiven if you think it paradoxical that the proto-black-hole died and collapsed *after* losing most of its mass to its companion. This common confusion results from a failure of many commentators to distinguish adequately between the initial mass of a star, which determines how fast its interior ages, and the composition and structure of the star's core, which contains only a fraction of its mass. It is the latter that determines the star's immediate fate, and in the case of Cygnus X-1, the proto-black-hole's interior had aged too far to be rejuvenated merely by losing weight. And a good thing, too, for if the proto-black-hole had collapsed (very possibly with an explosive release of energy in its outer layers) before losing most of its matter to its neighbor, then we would not have had a binary system left to study today. The recoil would have sent the black hole careening off into space, bereft of its brother and left to starve alone.

I synchronized my approach with the rotation of the system as it swung around its center of gravity. It was obvious that the star that had not become the black hole had aged considerably since its weight gain under dramatic circumstances. This was to be expected, because it was now quite a massive star. It was beginning to bloat, as more and more of its nuclear fuel was expended in producing its now-prodigious luminosity. It also looked far from normal, even for its age and mass. Because the black hole tugged more strongly on the hemisphere facing it, that entire side of the star was drawn into a vast, ungainly bulge, tapering almost to a point. I recalled the tides I had seen risen on the stars approaching collision in the Milky Way's center and the more dramatic distention that preceded the destruction of the star that had ventured too close to the black hole in my daydream. But unlike those doomed bodies, this pointy bulge was not growing with time. It seemed to have reached some kind of stasis, with matter from the star flowing into the distension and then spilling through the point. This black hole, it appeared, was not tearing its companion star apart but instead was subjecting it to a more exquisite, drawn-out form of torture.

The stream of gas being sucked off the star was not going quietly. It was full of turbulent curlicues, swirls, and sprays, with an occasional sudden discharge of gas that reminded me of a colicky water faucet. I tried to focus on individual eddies—swirling, dark spots in the turbulent gaseous fluid—and to follow them as they passed through the constriction and toward the black hole. This was not easy; the contrasts were subtle and the flow complex and unsteady. But I was able to discern some regularity in the flow and saw that the matter did not fall straight toward the black hole once it had passed the constriction. On the side of the constriction toward the black hole, the flow was strongly deflected from making a beeline for the hole. I supposed that it retained the memory of its original orbital motion, which carried it so far forward that it overshot its mark. Despite the exotic setting, I recognized this as the same phenomenon that kept satellites from falling to Earth, and the name of this orbital memory, "angular momentum," made it all seem commonplace. In a moment of panic, I feared that none of the matter spilling through the constriction would actually reach the black hole. Was this black hole, too, going to starve, despite being surrounded by a cornucopia of gaseous food? Then I remembered all those X-rays—obviously, *something* was feeding this black hole—and I pressed on.

6

Black Hole with a Mission

I followed the deflected stream as it gradually spiraled toward the hole. Now it was quite narrow and hence easier to follow, but I began to notice an increasing amount of swirling matter that was not part of the stream. The motion of this swirling gas was directed nearly at right angles to the straight path into the hole, and it did not diffuse through the entire space, as the tenuous corona surrounding the Galactic Center black hole had done. Rather, it seemed to be confined to a platter, or disk, hugging the plane in which the binary twins gyrated. The stream had to plow through this material, and eventually the resistance it encountered took its toll. The stream began to shred around the margins, to wobble in its path, and to spread from its well-focused trajectory. But it faced a still more formidable obstacle. About two-thirds of the way from the constriction to the black hole, the spiraling stream finished one complete circuit and crashed into itself. I quickly pulled away from the disk as I saw the great splatter of superheated gas looming ahead. Gas was thrown up out of the disk, and ripples spread out sideways as the swirling platter tried to come to terms with this unexpected disturbance.

This was the end of the stream—but the source of the disk. I watched the debris of the collision rock back and forth as it settled down, and I saw it spreading into a ring, which then broadened both toward and away from the hole. It thinned as it spread, until it merged into the very disk-like structure that I had encountered on the way in. The pattern behind its motion was now clear: The gas was distributing itself into nice, regular orbits about the black hole, just like a planet orbiting the Sun or a satellite orbiting the Earth. I thought of the Milky Way's disk, and if I imagined each star as an atom of gas, I could perceive a similar pattern. Just as the stars marched nearly in lockstep about the Galaxy's center of mass, so would the atoms of gas in the disk trace broad, stable circles about the black hole, oblivious to the danger they would face if they ventured too close. Because the disk was so much less massive than the hole, there was not even the risk that something akin to spiral arms or tumbling stellar bars would mar the regimental parade, as I had seen them do in the Galaxy. I knew that the gas could stay in that state of motion virtually forever. But clearly that could not be what was really happening, if gas was draining into the hole at the prodigious rate indicated by the X-rays. In order for it to spiral in, something had to be dragging on the orbiting gas, drawing angular momentum from it as it circled the hole. To find out what this viscous agent was, I had to enter the disk itself.

Did I dare to immerse *Rocinante* in the disk? It looked formidable, completely opaque, and not at all calm. Arches of luminescent gas kept erupting from beneath a roiling gaseous sea. Plugs of vapor would shoot out of the disk to various heights, spread out in striations along an arch, and then drain back toward the surface. Sometimes the arches would twist, merge, separate, or suddenly open up to the sky, lobbing their contents into space. The flitting network of arches looked like a pattern of iron filings created by a convention of drunken magnets, and even more like the prominences a colleague once showed me—through a specially designed telescope—erupting from the surface of the Sun. The latter flares were in fact magnetic in origin, as my colleague had found by measuring the polarized light (a sign of the helical gyration of electrons along the magnetic field).

The flares were obviously pumping out a lot of energy, because it was getting hot even before I reached the surface of the disk. The disk, it seemed, had an atmosphere of hot gas, presumably evaporated from the surface by the flares. I held my breath and descended through the surface. There was less resistance than I expected—the "surface" was more an impediment to vision than a physical boundary, although I could feel increasing discomfort as *Rocinante* was buffeted by the highly turbulent gas in the denser layers of the disk. The magnetic field was in evidence everywhere and stronger now than ever; lines of force and the striations they cause seemed to dominate the organization of the gas. Here they formed no arches, however, and they did not seem to be seeking the surface of the disk and freedom. They were tightly confined, combed out by the swirling orbital motion of the disk, much like marsh grass bent over flat in a raging flood.

The symbiosis between the magnetic field and the orbital motion—*that* was the key to how matter was able to spiral toward the black hole. I struggled to remember the theory that described the complex behavior of hot matter in the presence of magnetism. This subject had been my nemesis as a student. Magnetism, it turned out, gave hot matter an added elasticity and an odd kind of tension. It imbued ordinary gas with certain qualities of both rubber bands and the mainsprings of old-fashioned watches. Conversely, hot matter was able to grab onto magnetic lines of force and to stretch, twist, and bend them as though they were candy stripes frozen into toffee. This stretching was what combed out the field lines here, and the symbiotic backreaction of the magnetic force was what was dragging on the gas, allowing it to spiral in.

Confident now that matter would flow all the way in, I surfed along the disk toward the black hole. I dipped in and out of the obscured disk interior, enjoying increased confidence in my mastery of *Rocinante*'s controls. The speed of the swirling gas increased, as predicted by the theory of gravity, and so did the temperature, the luminous glow inside the disk, and the violence of the stretching and churning magnetic fields both inside and out. The disk's radiance grew so strong that it seemed to be bucking the dips in my inward motion, pushing me away from the disk's central plane. This was probably an illusion, but it reminded me that radiation itself, when intense enough, can actually levitate matter against the pull of gravity.

I fully intended to reach the place in the center of the disk where matter departed from its gradual spiral path and made its final plunge into the hole—the "orbit of instability" that I had so carefully avoided in the center of the Milky Way. That place, I calculated, should be about 100 kilometers from the horizon of the black hole, which turned out to have a mass equal to that of 15 Suns. I was now so adept at navigating *Rocinante* that I would have felt comfortable going in nearly that close. Noting that the stream flowing in from the companion had merged into the disk at a distance of about 1 million kilometers from the hole, I relaxed and watched the kilometers tick off as I sped inward. Then, suddenly, I became violently ill. I had never been so sick in my life. I checked my distance from the hole: 10,000 kilometers. What could be happening? It felt as though my torso and upper body were being pulled away from my lower extremities. I was riding with my head toward the black hole and was able to achieve a minor increase in comfort by turning my body sideways. But the respite proved temporary. My craft was on autopilot, heading straight for the hole, and I soon felt as though my head were being pulled apart, front from back. And as for my gut . . . well, I will spare you the details.

I'm sure that I was about to pass out, and it was sheer luck that I managed—with a great effort of will—to locate and activate the reverse lever. I pulled back out to 100,000 kilometers, up and away from the disk, and collapsed in a cold sweat. The disk formed an immense floor in my field of view, its center a barely discernible spot in the distance. As I regained my composure, I immediately realized what had happened to me. At first, the black hole's gravity had not posed a problem, because I had been allowing *Rocinante* to fall almost freely toward the hole. Like any astronaut orbiting Earth, I had felt virtually weightless. But my head, being slightly closer to the black hole than my feet, had actually been subjected to a slightly stronger gravitational pull, and that *difference* is what had gotten me. When I came within 10,000 kilometers of the hole, the difference in gravitational pull had approached and then exceeded the normal gravity of Earth, under which a human body is designed to operate. I was indeed being pulled apart. It was as though I had somehow discovered a way to stand upright and upside down *simultaneously!* The blood (and every other fluid in my body) began to rush away from my middle and to pool in both extremities. A force comparable to my weight on Earth was stretching every tissue in my body. You can imagine the discomfort. When I turned sideways, the effect was diminished: Because I am thinner than I am long, the difference in the gravitational pull was correspondingly reduced. But as I approached still closer to the black hole, even the difference across my torso became intolerable. In a sense, I had suffered a milder version of the fate that befell Cygnus X-1's companion, whose midsection had been extruded into a pointy bulge from which the accretion stream was pulled toward the black hole. I thanked the deity of mass transfer in binary stars that my encounter with extreme tidal forces had not proceeded to the gruesome outcome that was keeping Cygnus X-1 fed.

Apparently, I was not to close in on the black hole after all. I was up against physical laws that I could neither change nor circumvent. Fortunately I am a stoic at heart, and I knew that remote observation via telescope, though not the ideal way to collect data, sometimes has to do. One simply accepts that limitation as an astronomer. My ego needed to be reminded that, after all, I was closer to a black hole than anyone else had ever been. I could live with the disappointment of not getting quite as close as 100 kilometers.

Having come to terms with this unforeseen restriction on my movement, I surveyed the disk below and ahead of me. From this height the magnetic flares seemed less significant, little more than minor viscous eruptions like the bubbles and cheese-spouts that erupt from the surface of a simmering fondue pot or, given the disk's thinness, a bubbling pizza (my harrowing encounter with the black hole's gravity had unaccountably given me an appetite). I could see the underlying large-scale structure of the disk, stretching away toward the black hole. The disk was luminous everywhere, and the more so the closer one got to the hole. With increased radiance came the expected increase in temperature, manifested in the progression of "colors." I have to put this word in quotation marks, because the progression far surpassed the sequence from red to yellow to blue-white that a human eye perceives when a metal bar is heated to thousands of degrees. Here the colors went off the human scale at violet proceeding through the ultraviolet, far ultraviolet, extreme ultraviolet, and on into the X-rays. Apart from these gradations of temperature and radiance, the disk was overwhelming in its flatness and featurelessness.

To get my bearings, I focused on the hottest and most luminous region of all, the area surrounding the black hole. I could see the blackness in the center and the disk continuing beyond. There the flatness appeared to be broken by a curious asymmetry. Just the other side of the hole, the disk seemed to warp upward, forming a distorted band that folded over on top of the gap that signified the location of the black hole. I moved around toward the far side to get a better look, but the distorted region of the disk seemed to compensate for my motions, always remaining exactly on the opposite side of the hole. It dawned on me that I was witnessing an illusion caused by the wrapping of light rays around the black hole as they crossed the intense gravitational field on their way to my instruments. I should have expected this. I knew that radiation, as well as matter, responds to gravity, though it was shocking to see the effect so forcefully displayed. There was another asymmetry as well, but the origin of this one was easier to visualize. I noticed that the disk was a little bit brighter, the X-ray luminescence just a little bit harsher, to the right of the hole than to the left. This, I immediately recognized, must be due to the "headlight effect," the forward beaming of the radiation emitted by moving matter. In this case, the light would be beamed toward the direction of the disk's spinning motion, and the brighter side had to be the side that was coming toward me. According to the theory of relativity, an asymmetry this pronounced could mean only one thing: The matter that I was viewing swirled around the black hole at nearly the speed of light. These were all well-known theoretical predictions, but by this point I was so awed by the spectacle that I was disinclined to trust the intuition I had developed through years of painstaking study. As if to pinch myself mentally, I double-checked that the disk was spinning in a clockwise direction, as it would have to be if light from the right-hand side were really boosted by its oncoming motion. It was.

I tried to follow the gas as it disappeared through the black hole's horizon. The horizon was a sphere with a radius of 45 kilometers. Was I right that the matter in the disk would break its slow inward spiral and plummet into the black hole from 100 kilometers farther out? I saw what looked like evidence of a dramatic change in the disk's state of motion near the predicted place, but I couldn't be sure; the hot atmosphere above the innermost parts of the disk became so thick that it all but obscured my view. At the other end of the death-plunge I could barely make out the gas fading just outside the horizon. The textbooks appeared to be right. The X-ray radiation regressed through the colors as it fought its way out through an increasingly debilitating gravitational field. As the signal faded, the color passed back through the varieties of ultraviolet, blue, yellow, and red, and then it was too faint to tell. What happened in the intervening space between the horizon and the orbit of instability? The artists' impressions in my musty old textbooks had represented the disk as having a distinct inner edge, but this seemed unlikely. This disk was so rich with matter, right to the center, that it could not have thinned out enough to become transparent. I sighed with relief at this confirmation of my "theorist's instincts." Still, I felt one last pang of regret that I would be forever denied a closer view.

It could be that I was fortunate to have been forced to take a long view of the disk, for I began to perceive certain causal connections that had eluded me before. The black hole's gravity caused motion, first the open spiral of the narrow accretion stream, pulled from the companion star into the black hole's sphere of influence. I looked back, away from the black hole, and saw the distant panorama of the stream swinging one full turn around the system and crashing into itself, that big splash now dwarfed by the immensity of the disk. I saw the incipient disk spread away from the crashing stream, both inward and outward. Most of the gas spiraled inward and, as it did so, drifted ever deeper into the gravitational pull of the black hole, where it swirled faster and faster while its increasingly ferocious magnetic fields (stretched and amplified by the motion) tugged at it chaotically until it heated up. Thus motion begat heat, and heat begat luminosity. Four trillion tons of its companion's substance was disappearing into the black hole of Cygnus X-1 every second, but 100,000 times the luminosity of the Sun was coming out. It seemed a fair trade. This black hole *was* an efficient engine for turning matter into energy, an intense flickering source of X-rays that finally reached Earth.

I pulled out my measurements of the Milky Way's central black hole, expecting to find that it was starved of nearly all matter. That would surely explain why its environs emitted just the barest of glows rather than blazing away like Cygnus X-1. But the Milky Way's black hole seemed not to be playing by the same rules. Despite its sparse surroundings, that huge black hole was actually swallowing matter more rapidly than Cygnus X-1 was gobbling its companion. Somehow the chain—matter plus gravity goes to motion to heat to radiation—had been broken, the trade of matter for luminosity not consummated. The bigger black hole was greedy, swallowing most of the heat along with the matter, before the heat could turn into luminosity and radiate into space. I pondered an old theoretical idea in the hope that it could explain such a difference. The Milky Way's central black hole was 100,000 times heavier than Cygnus X-1. *Its* horizon was millions of kilometers across. I focused on this idea of relative size and what it might signify. The amount of matter flowing toward the black hole in the Galactic Center was indeed modestly larger, but the dimensions over which this matter was forced to spread were *vastly* larger. Could this be the clue I was looking for? Spread out so thinly, perhaps the matter never reached the concentrations at which heat could be generated and released efficiently.

It sounded like a plausible theory, and it gave me that temporary glow of self-satisfaction that theorists sometimes get just before delving further into an idea and realizing that it is much more complicated than it had seemed at first. True or not, I couldn't think of any way to test this hypothesis without observing many more systems than just these two. I was more confident in my ability to explain what I had seen while snooping around the environs of Cygnus X-1. At least that seemed to make sense. Matter is drawn inward, and luminosity flows out in proportion to the matter flowing in. I decided to concentrate only on black holes with donor stars and orderly disks. Then I noticed something that threatened to unravel even this cozy fragment of my story.

7

SS 433

The problem was that not all of the matter approaching Cygnus X-1 made it into the black hole. I should have seen this coming, although in my defense, I must point out that I was looking for evidence of inflow, not outflow. I had already noted that a few wisps of gas managed to escape, expelled by the dynamic action of magnetic flares. A little evaporation here and there was no cause for alarm, merely another example of the normal wastage that always seems to accompany physical processes in the Universe. But these filaments were not isolated escapees. Together they wove a disturbing pattern in front of me as I peered toward the distant black hole. Only as I retreated, and after I had spent some time examining images from my visit, did I see how much of the gas in the disk departed from the inward drift of the great viscous spiral. There it was, a streamer of gas shooting away from the disk, and there another—several streamers at once, twisting together, then merging and finally forming a jet of matter screaming off at high speed, perpendicular to the disk. Could it be that black holes expelled matter as well as drawing it in?

Of course, I knew of a system that was an extreme example of this phenomenon. SS 433 was its surprisingly unromantic name, given how ingrained it had become in my generation's store of iconic imagery. I hadn't thought about it recently, but suddenly it seemed to pose a towering challenge to my developing world-view. The intensively gravitating body of SS 433 seemed to be expelling far more matter than it absorbed.

Everyone in my generation of scientists remembered the hoopla that surrounded the discovery of this object and its peculiar properties. It was one of the signal events of my early astronomical education. The fact that it ultimately proved to be representative of a very rare class of beasts may have knocked it out of the pantheon of astrophysical archetypes (few would rank it with pulsars or even with those mysterious bursts of gamma-ray radiation that were being so hotly debated at the time I left), but SS 433's discovery had the element of surprise that reminds one that one's view of the Universe may have to be revised on short notice.

I will never forget the first announcement, because I missed it. This was at one of those conference series that are named not for the place at which they are actually being held but for the place at which they were first held. I was attending a meeting associated with one such venerable institution—the famous Texas Symposium on Relativistic Astrophysics—in Munich in 1978. It was late in the week, and my attention was flagging. The lectures were dull, and little new was being reported, so I took the afternoon off to visit one of the art galleries for which the city was famous.

This was a big mistake. By the time I had slipped back to the enormous hotel ballroom in which the lectures were being held (in order to be seen dutifully nodding off during the concluding session), the buzz was just dying down, and all the astrophysicists were dashing off to the airport to report the new discovery to their groups. (Quaint world when there was no Internet!) Everyone was in such a hurry that I couldn't get a straight story of what had happened, but I caught murmurs about "Doppler" and "anomalous redshift." I slunk back to my home department, and there I encountered, for the first time as an insider in the profession, the kind of hyped-up astrophysical press coverage with which we all subsequently became familiar. Reports of the discovery were featured under multicolumn headlines in all the papers and newsmagazines. "Mystery Star Both Coming and Going" blazed a typical headline. As though eager to avert panic, venerable TV newsreaders relayed the opinion of astronomers that "this is some kind of star that's in some terribly weird kind of trouble." Like everyone else, we cobbled together the rumors and fragmentary observational reports that had come in since the first announcement and began work on our own theories, in order not to miss out on a possible "scoop."

Rocinante signaled an abrupt heads-up from my nostalgic reverie. I was already heading into the environs of SS 433. There was no doubt that matter was being powerfully expelled from this system. Tens of light-years away from the destination itself, I encountered the vanguard of its effects on the surrounding regions. Like the bubbles around luminous stars that I had seen at various locations en route to the Galactic Center, SS 433 had inflated a hot cavity around itself, but with a difference. This cavity, instead of being spherical, sported a pair of highly elongated protrusions. The jets were pumping their energies and momenta in two diametrically opposed directions, boring into the interstellar matter like the high-pressure water jets sometimes used to excavate mine shafts. Like those pulsing water streams, the jets plashed vigorously against the working surface and, having spent only a fraction of their force, ricocheted into the main cavity, broadening it as an afterthought. I felt a minor jolt as I crossed the sharp boundary from undisturbed interstellar space into the pressurized cavity. As it happened, my route took me across the path of one of the jets, which was by this point much more diffuse and spread out than when it had left its source; nevertheless, there was no mistaking its impulse. With difficulty, I made out the pale, multicolor luminescence it left in its wake (here X-ray bright, there a mixed pinkish and ultraviolet glow from disturbed hydrogen).

The scene was grand, but there were no surprises here. I recalled the pictures I had seen of W50, the SS 433 cavity with "ears," long before my visit. Identification of the ears with the impacts of the jets had been one of the comforting verifications that had tied the whole picture so neatly together. But that had come well after the first flush of discovery. I thought back to the observation that had triggered the initial excitement, inspiring my colleagues' hushed words about "abnormal Doppler shifts" and the like. I will not insult the quality of your liberal arts education with yet another disquisition on how the Doppler effect works. My style manual indicates emphatically that a discussion of the Doppler effect, or at least a description of its manifestations and uses, should appear in the first chapter of any astronomical memoir. I am perhaps treading dangerously close to new stylistic territory by not having informed you before Chapter 7 that I intend to dispense with this formality. If the mere squeal of an approaching train whistle or groan of a receding police siren does not elicit the expected Pavlovian response (thoughts of distorted ripples on a lake; mental images of line drawings in physics textbooks, with Pepto-Bismol-pink backgrounds), you may wish to consult Chapter 1 of any number of available works.

What I will describe is the dramatic (and unexpected) role that the Doppler effect played in the discovery of SS 433's true nature. The designation itself connotes nothing extraordinary. The intention of the cataloguers (Messrs. Stephenson and Sanduleak) had been to record stars that were unusual only in their production of intense spectral "lines," blips of extra emission at certain very precise colors. These lines are the products of electrons dropping between equally precise orbits in their respective atoms. For these lines to appear with intensity requires special but not exceptional conditions, and the available colors themselves are the well-documented properties of the chemical elements in various states of excitation and disarray. These intrinsic colors are not to be fiddled with.

Number 433 in the Stephenson and Sanduleak catalogue had a surprise in store. When measured accurately, the lines had the wrong colors. More frighteningly, the colors changed with time. Enter the Doppler effect, according to which these well-recorded lines are expected to have their original hues only if the luminous matter is at rest. If it is moving in the direction of the observer, the color will "squeal" toward bluer hues; if away, it will "groan" and turn redder. SS 433 showed three sets of lines—one groaning, one squealing, and one more or less in repose—and therefore seemed to exhibit three states of motion at once. As the headlines said (do they ever lie?), SS 433 was indeed both coming and going, not to mention standing still, and clearly had to be in some terrible sort of trouble.

Only it wasn't really in so much trouble. Like the ending of a classical comedy, the resolution of the paradox showed that all was right with the world. The change of hues with time, which (in wilder moments of speculation) had seemed to threaten the foundations of physics, proved to be their salvation. After the lines had been monitored for a few months, a simple pattern emerged. The approach and recession of the "moving lines" oscillated, changing places twice every 164 days. A simple model fit the data perfectly: SS 433 was spraying out a pair of jets in opposite directions, and these rotated with the aforementioned period. The rotation of each jet was a kind of precession, like a searchlight plumbing the sky 20 degrees off the zenith. An early analogy to a rotary lawn sprinkler also proved apt, for analysis of the lines revealed that matter in the jets did not stream out smoothly but rather chugged out impulsively in bullets or blobs of fluid, several per day. Twice per rotation period, the matter in the jets would be moving perpendicular to our line of sight and, according to the classical theory, should show no Doppler shift. But when these times came, the hues were still shifted to the red. Even this puzzle was quickly resolved. Long ago, the theory of relativity had predicted that time would slow down for a body in motion. Even though it was moving neither toward nor away from us, the matter in the jets was moving, and this dilation of time showed up, exactly as predicted, through a redshift—a slowing down of the frequency of light. Another prediction of Einstein's theory was confirmed, the steel trap of its validity tightened. And the speed of the matter in the jets could be deduced with precision. It was a quarter of the speed of light.

While I ran through all this history in my mind, I was steadily approaching SS 433. The smeared-out jet that I had crossed as it plowed through the outer reaches of W50 now seemed to be resolving itself into much finer structure. Each bullet was emitted in a slightly different direction, as the sprinkler head chugged around, and thereafter traveled in a straight line. Far enough away from SS 433, the collective pattern of all these bullets (if they survived) should delineate the surface of a hollow cone. As I crossed back into the jet's path, I found that the impulse was no longer spread widely but was confined to a conical surface, spread out by about 20 degrees from its axis of symmetry—just as predicted. Closer in, I could see the helical pattern traced by the sequence of bullets emitted during successive rotations, a pattern that was washed out when viewed on large scales. I recalled how astounded I had been when my radio astronomer colleagues had imaged this very helical pattern, all the way from the 17,000-light-year distance of Earth.

It was all very familiar, the geometry and motions of the SS 433 jets, and yet the congruence between what I saw and what I expected made it spooky. This level of predictability and economy of assumptions was seldom attainable in models of real astrophysical systems. Aesthetics had always been a principal goal in my approach to theoretical astrophysics, but I had grown used to regular frustration of my hopes to attain this ideal. Astronomy as practiced was nearly always about data, which seemed without fail to sneak in a twist, a complication, an exception to every rule. The systems we studied were so complicated that elegant models seldom worked as well as they did in the case of SS 433. Maybe this, even more than the simple presence of its steadily precessing jets, accounted for the iconic power that SS 433 seemed to have over my colleagues and me.

In any case, this familiarity was soon to be shattered. The theory of SS 433 was cut and dried only insofar as its outward trappings were concerned. My present objective was to find the motive force of its jets. Here, where it counted, I ran into one obstacle after another.

From my present vantage point, I could see the basic structure of the binary system, which was not all that different from Cygnus X-1 in some respects. As in Cygnus, the star was being sucked away. It looked like a similar kind of star, too, one of those blindingly blue ones whose ultraviolet rays could give you a sunburn far worse than Earth's star, even if you had the protection of a pea-soup atmosphere. (*Rocinante*, fortunately, had far better armor than that.) SS 433 and its companion were swinging around each other once every 13 days instead of every 5 1/2 days, but this difference in detail seemed hardly worth mentioning. What *was* different was what appeared at first to be a general haze enveloping the SS 433 system. This, I saw, was actually a powerful wind rushing away from the system at speeds up to 2000 kilometers per second—70 times faster than Earth's march around the Sun and 8000 times faster than a jet plane. The flowing gas glowed softly in X-rays, punctuated by arcs of extreme X-ray intensity and outward-rushing spots from which emanated the cool pink glow of hydrogen accented by yellow tints of helium. The arcs were telltale signs that this wind was violently turbulent and unsteady; X-rays were produced where the faster streams plowed full-force into more slowly moving matter. But what was the wind rushing away from? Was the wind coming off the accretion disk around SS 433, or off the star that was feeding it, or both? I couldn't tell. The cool clouds reminded me of debris caught up in a gale, like the blowing palm fronds and bits of plywood that had once terrified me when I was caught in a hurricane. But what was the source of the debris? Which part of SS 433's binary system was being shredded and blown away by these gusts?

Through the light fog I could see the stream of gas crossing the gap between the tortured star and the accretion disk. It was more ragged and active than the stream in Cygnus X-1, and it looked like some of the debris—though not all—was being torn off its margins. Compared to this gushing firehose, the stream feeding the disk in Cygnus X-1 looked tame; thus it came as no surprise to me when I later deduced that matter was flowing across to SS 433's cauldron at 100 times the rate at which mass was being transferred to Cygnus X-1. I peered ahead to see whether I could spot the hole at the center of the disk, if indeed there was a black hole there, but all I saw was an X-ray/ultraviolet shimmer behind a gauzy screen.

With difficulty I descended to trace the lay of the disk and saw that, unlike the swirling platter in Cygnus X-1, it was not flat. A rolling warp, its spiral twist sweeping one turn around the center and coming out to envelop *Rocinante* as though in the trough of a gigantic sea swell, disoriented me and made me slightly woozy. The disk was apparently wobbling like a top, guiding the 164-day precession march of the jets, but for now the reasons eluded me. I cautiously crept inward, but the roller-coaster scene had triggered a panic attack that I had difficulty controlling. What if I should encounter those sickening tidal forces again? Still I inched toward the center of the disk, jaw clenched and palms moist. At 10,000 kilometers from the center I felt the first twinges of tidal stretching—and froze. I could go no farther.

Neither could I see any farther. The churning gas just in front of me became so dense that it formed a glowing wall, its outer boundary so indistinct that it seemed diffuse and opaque at the same time. There was no disk from here on in. What had been a thin orbiting structure (its warp superimposed on it like the warp in a potato chip) had ballooned into an immense quasi-spherical bulge, as though some force from within were trying to blow it apart. As if to underline its dispersive tendencies, this gaseous globe expelled a good fraction of the powerful wind and debris that I had puzzled over earlier. But the force I was familiar with, gravity, could not be doing this: Gravity was always attractive! The bulge hid the center from further scrutiny, and I had to pull up and creep along the diffuse surface to avoid flying blind. I noted that the bulge still retained a great deal of the disk's rotation and thought that maybe, if I headed for the rotation axis, the flow would open up into an evacuated funnel, like a whirlpool in the sea or (more prosaically) in the water going down a drain. But like the maelstroms that wreck ships in classical sagas, this one had its own version of a waterspout shooting up through the center: the jet. As I came over the lip and one of the rotational poles came into view, I was blown away (almost literally) by the spectacle of this massive ejection. The jet wasn't emerging calmly through an evacuated funnel lined with smoothly rotating gas; it was blasting its way through with a great deal of violence. True, the centrifugal force of rotation created a preferred alignment for the jet's path, but once it had found the weak spot, the jet forced the gas aside, opening up a clear channel by its sheer impact. Through this channel a searchlight beam of X-rays and ultraviolet rays accompanied the jet out into open space, and the matter in the jet fluoresced in the glow of its own sheath of radiation. Even the protons and neutrons inside the nuclei of some of the atoms were disturbed by this tremendous agitation and emitted a spectrum of gamma rays. With atoms knocking together at a quarter of the speed of light, this was not hard to comprehend.

I tallied up the matter flying away from this object, in all its forms—hot wind, debris, fast jets—and came to a startling conclusion. SS 433's intense gravitation was extracting from its companion 100 times more matter every second than was disappearing into Cygnus X-1's maw. But instead of swallowing this matter, SS 433 was shooting most of it back out into space. Why?

I saw how such an imbalance *might* come about. As matter sinks deep into the gravitational field, it acquires a lot of energy. First comes motion, as the matter is accelerated by the gravitational pull, then perhaps heat, and then radiation. I had already worked out this chain of energy flow for Cygnus X-1. But I hadn't taken into account the many ways in which energy can be transported from one place to another. Heat is conducted from high temperature to low; radiation leaks out into dark space; stretchy coils of magnetic field set widely separated gaseous filaments in motion; hot cells of turbulent fluid boil up through a cooler atmosphere. In each case, energy is deposited somewhere other than the place in which it was liberated, and deposition of energy can lead to motion where it is least expected. The pressure of radiation forcing its way through from below was perpetually blowing apart the shroud of SS 433, puffing it up from within until it released its wind and debris. I was less certain what mechanism propelled the jets, because their origin was hidden—but I now understood that means, opportunity, and perhaps motive could readily exist for this seemingly illegal escape from the clutches of a black hole. Whatever the mechanism, it wouldn't take much for a small amount of matter, skimming just outside the black hole, to unleash enough power to blow away a much larger amount of matter that never even got close.

This all seemed to make sense, but did it require that there actually be a black hole at the center of SS 433? I had read somewhere that a debate still raged over whether the enshrouded collapsed star was a neutron star or a black hole, because attempts to estimate its mass had been inconclusive. I suppose I could have solved that problem on the spot with a few measurements of speed and distance for my freely falling craft or with precise orbital measurements of the two stellar companions. But I was too preoccupied by what I had just learned to worry about such refinements. Black hole or neutron star, I doubted that it would make much difference to the outcome. A neutron star's gravity had to be nearly as strong as that of a black hole, and it was the competition between gravity and motion, not any special properties of the gravitating body, that was in question.

What I now understood was that gravity and motion were not always so finely tuned. I had seen some examples of an exquisite balance—were they the exception? My thoughts returned to the steady march of the stars orbiting the Galaxy, in nearly perfect circular orbits. Even the slight imbalances that led to spiral arms—in which gravity wins a little bit each time a star slows down as it encounters one of those interstellar traffic jams—had a certain grace and delicacy about them. The tidal disruption of a star by a black hole was more heavy-handed and really quite violent, but at least it was a fleeting event that occurred and then was over. Except for a modest escape of some of the accreting gas, Cygnus X-1 made a clean trade of matter for energy. Yet the violent disequilibrium of SS 433 seemed to be a chronic condition. Did the black hole (or neutron star) not know its own strength? Did it grasp the substance it sought to incorporate with such vehemence that it overshot and flung most of it away?

I knew that gravity and motion were not really out of control. All of this had to be predictable. It was just shocking to see matter flout such a strong gravitational field. And SS 433 was not unique. There were episodic jet-emitting black holes that produced surges every so often, for reasons unknown. In these cases, the entire insides of a disk would suddenly drain toward the black hole (like the extremely low tide that precedes a tsunami), only to resurge in the form of jets shooting outward along the rotational axis at speeds even closer to the speed of light than SS 433's jets. Apparently, black holes did not exist simply to attract and devour, as popular literature would have it. Surely this would be the impulse of gravity, left to its own devices. But I saw that when combined with the curious and often contrary properties of matter—its momentum, pressure, radiation, and magnetic fields—gravity could often repel matter or expel it, or even compel it to whirl with wild abandon. Yet in the disk of the Milky Way, the departures from regularity were subtle. Somehow gravity was able to play with motion to achieve a nearly perfect balance. I knew, of course, that the same thing, on a microscopic scale, prevented the Sun and all stars from collapsing in on themselves, at least for a while.

I had a choice to make. Should I seek out examples of ever more violent and contrary phenomena driven by gravitation? I felt naturally attracted to the dramatic, the surprising, and the exotic, and at first this seemed to argue for undertaking the quest after super-fast jets. Probably I was also attracted to the idea that jets mounted a heroic resistance to authority: the victory of the underdog in standing up to gravity. But there was also a kind of tension and high drama in the challenge I finally selected—the quest for equilibrium. For I knew that equilibrium was not just stasis; it was a standoff in the fierce competition between huge forces of attraction and equally huge forces of resistance. When expressed in these terms, even the benign equilibrium of the smiling Sun seemed to survive in a kind of "balance of terror." One uncompensated deviation on either side of the fulcrum could lead to disaster.

As I debated with myself, I began to pull away from the SS 433 system, not really sure of my next destination. One more feature of SS 433 nagged at me. What caused the disk to warp? This really didn't seem to make sense. The matter flowing across from SS 433's companion star arrived imprinted with the spin of the binary's orbit. If anything should have been a gyroscope, retaining its orientation as the world around it turned topsy-turvy, it is that disk. Could this too be a symptom of the general failure to attain equilibrium, with a portion of that redeposited energy—some radiative, magnetic, or thermal torque—twisting the disk away from its preferred direction? Remarkably, even that seemed possible. And it also seemed possible in Cygnus X-1, whose disk I had perceived as being so flat. Hadn't I read somewhere that it, too, showed some evidence of a wobble? Maybe, in my naïveté, I had wished it flat. Knowing now that the interactions between gravity and matter could be more complex than I had ever imagined, I turned my attention to the next phase of my exploration, the quest for a rock-steady truce between gravity and matter.

Part Three

Equilibrium

8

Shangri-La

It struck me, as I accelerated away from SS 433 and prepared for hibernation, that more than 65,000 years had passed on Earth since my departure. In all likelihood, the scientific conundrums I was confronting—so far, with mixed success—had long ago been resolved by more ingenious if less direct methods . . . assuming that science was still practiced at home. The goals, norms, and technologies of civilization must have been altered far beyond recognition by now. How many other travelers, or colonies of travelers, from Earth might now be sharing interstellar space with me? Or had physical space travel been a transient fashion, quickly superseded by something more efficient, perhaps the transmission of some digital "essence" of intelligence and personality without the need for an accompanying organic receptacle?

The passage of time might be quite irrelevant to such a virtual being. I, on the other hand, had to worry about aging. As I intimated earlier, I address this problem by manipulating the passage of my time. There is nothing mysterious about this. I exploit the most elementary consequence of Einstein's special theory of relativity, the effect known as time dilation. As viewed from the Earth or from nearly any other observing platform in the Galaxy, time flows more slowly for me, simply by virtue of my high speed. The astute reader might have noticed my allusion to the phenomenon of time dilation when I explained how SS 433's motions had been deduced using Doppler shifts. To show why time dilation occurs and how to exploit it, I first need to explain a little bit more about the principles of relativity.

People had known since the seventeenth century that light travels with a finite speed, about 300,000 kilometers per second. It had seemed self-evident that if you could travel at, say, 150,000 kilometers per second toward the source of a light beam, then you should measure its speed as 300,000 + 150,000 = 450,000 kilometers per second. After all, in normal experience speed is relative: When you pass a car on the freeway, it actually seems to be going backwards, relative to you. However, in 1905 Albert Einstein realized that speeds close to the speed of light behave differently. For example, if a truck is barreling toward you at 200,000 kilometers per second (two-thirds the speed of light) and you are suicidal and head toward it at 100,000 km/s (one-third the speed of light), then the two speeds do not add up to the speed of light, as a simple sum would suggest. Instead you will find yourself closing in on the truck at only $^{9}/_{11}$, or 82 percent, of the speed of light. This is because speeds become "less relative"—they do not simply add up—as they approach light speed. The speed of a light beam is not relative at all; it has exactly the same value, no matter how fast you are going and in what direction. What Einstein showed is that space and time, not speed, are the fundamentally relative quantities. Space and time become distorted—stretched or compressed—according to your state of motion, in such a way that all observers will agree on the value of the speed of light.

Time dilation is one of the more dramatic consequences of the principle of relativity, and it is entirely responsible for my ability to travel across the Galaxy within my natural lifetime. After my encounter with the black hole's tidal forces, I was especially sensitized to my own physiology, so let me try to illustrate this effect by using the example of my beating heart. My heart possesses a little clump of nerves to one side that sends an electrical signal wrapping around the entire muscle every second, telling it to beat. The chemical processes that set this timer are complex and involve molecular interactions that crisscross this natural pacemaker in every direction and at a variety of speeds. But I will simplify for the sake of argument and suppose that the billions of molecular reactions can be visualized as a single pinball that bounces around inside the pacemaker at the speed of light. The trajectory can be as fiendishly complicated as you like, provided only that the pinball hits the same bull's-eye once every second, triggering my heartbeat. You will have to trust Einstein's velocity transformation law to take care of deviations from the speed of light in the real pacemaker, and the law of averages to correct for my substitution of a single projectile for multiple concurrent interactions.

First look at the operation of this little synaptic device from my point of view. Between heartbeats the pinball, traveling at the speed of light, covers a distance totaling one light-second, or 300,000 kilometers. It must take a tortuous path indeed! Now let me revert to an observer watching my craft zoom past, from left to right. According to the principle of relativity, this observer must also see the pinball as moving at the speed of light, but now the heart is moving as well. When the pinball happens to be heading rightward, it has to chase the heart, which is moving along at nearly the same speed. These legs of the pinball's trajectory are therefore stretched and must take longer. The opposite is true when the pinball is moving toward the left: The heart then overtakes the pinball, and these legs are compressed. When the pinball is moving in any other direction, one gets a result between these extremes. What is amazing is that the total length of the ball's trajectory between beats, as viewed by the outside observer, is *completely* independent of the trajectory taken and is *always* longer than the trajectory as measured inside the spacecraft. Because light-seconds of distance translate into seconds of time, the outside observer sees my heart beating more slowly, and because any of my physiological processes may be represented by the same pinball analogy, to the outside universe I am aging more slowly.

How much more slowly depends on how close my speed is to the speed of light. When I am within 1 part in 100 of the speed of light, for example, I age 7 times more slowly. I call this slowdown in the passage of time my "Shangri-La factor." At 1 part in 1000 less than the speed of light, my Shangri-La factor is 22, and so forth. What I do in order to cover these vast distances is simply to accelerate, until I have reached the halfway point, at the comfortable (some would say leisurely) rate of 1g, the acceleration of any object dropped off the top of a building on Earth. Then I decelerate, at the same rate, for the remaining half of the distance. This has the advantage of giving me a sensation identical to the normal force of gravity (except when I choose to experience weightlessness during brief periods of free fall, as when I attempted to approach Cygnus X-1). I could have withstood somewhat higher accelerations, but one never quite gets used to them and they really are not necessary. At my leisurely rate and starting from rest, it takes just 2 1/2 years of my time to reach 99 percent of the speed of light. To reach the Galactic Center, I had to accelerate for 10 of my years (most of which were spent in hibernation) and then to decelerate for an equal time. My peak speed on this leg was within 1 part in 300 million of the speed of light, and (when I was awake) I was aging 13,000 times more slowly than everyone on Earth. In the 65,000 years that had passed on Earth since my travels began, little more than 60 years had elapsed inside *Rocinante*. Given my rigorous hibernation schedule, I had aged less than a decade!

9

The Soft-Shell Crab

Now I needed all the speed I could muster, not because I sensed an impending midlife crisis but rather because I wanted to get the next leg of my journey over with. For the first time, I was to travel beyond the circle of the Sun's orbit around the center of the Milky Way. Cygnus X-1 lay at roughly the same distance from the Galactic Center as the Solar System, but my detour to SS 433 had forced me to backtrack halfway to the center. Thus, although my next destination lay only 6000 light-years outbound along a continuation of the line connecting the Solar System to the center of the Galaxy, it was nearly 22,000 light-years distant from SS 433. With the time compression of the Shangri-La effect in force, I settled in for another 20 years en route.

Although I was hurtling toward a famous target—the Crab Nebula—I was not entirely thrilled with the prospect of visiting it. For one thing, I like my nebulae bold and extravagant. Orion, the Trifid, the Lagoon, Eta Carinae—those are my ideas of nebulae, sprawling fluorescent spectacles illuminated by newly formed hot stars. The compact, prolate fuzzball that I remembered as the Crab from my observing days on Earth was shot through with luminous red and green filaments, and a diffuse bluish background glow added a unique, if slightly garish, touch, but there were no bright stars to be seen. And with dimensions of barely 9 by 12 light-years across, this nebula was clearly a wimp in comparison to the others. In 1784 Charles Messier had placed it first on his list of objects that could fool comet hunters, a designation that seemed apropos—from a distance it resembled a comet that was still so far from the Sun that it had not yet sprouted a tail.

Moreover, to me the Crab Nebula looked nothing like a crab. It was Lord Rosse, constructor of the world's largest telescope in the 1840s, who had given it that label. It makes one wonder whether he had ever seen such a creature. I pulled up a copy of Rosse's earliest pencil sketch in my database, and to be honest, his own drawing looks more like a pineapple than a crab. (Curiously, his version of the nebula *does* have a tail, which reminded me of the pineapple's crown.) He later repudiated that early impression (made with a smaller telescope, not his "Leviathan of Parsonstown"), replacing it with a more familiar-looking mottled lozenge. Still no resemblance to a crab, despite the cage of thin filaments that was later made visible with the advent of photography. The nebula didn't even have an association of place, lying as it does (from Earth's perspective) near the tip of one of the Bull's horns, one full zodiacal notch away (clear the other side of Gemini!) from the arthropodous constellation of Cancer. And so far as I could remember, it had nothing resembling what I would call a shell—a shield of luminous or compressed gas, for instance. At best, it might be able to pass for "The Soft-Shell Crab Nebula." Or maybe the term *crab* refers to the curmudgeonly attitude with which I was approaching this system. Perhaps it had inexplicably been engendered in Galactic explorers of Victorian times, as well.

The Crab had to be my next destination, for one reason. Near its center lay a neutron star, the remnant of a star whose explosive demise had been witnessed on Earth in A.D. 1054. In theory, a neutron star resembles a black hole as closely as a body can without actually becoming one. Somehow its collapsing precursor knew how to stop gravity in its tracks, to create the tightest possible form of equilibrium. It weighed more than the Sun, but it was only 20 kilometers across. I needed to see what it could teach me.

Despite my misgivings about the visit, I took some consolation from the fact that the Crab Nebula, or its precursor, had once known true glory. Before A.D. 1054, no Earthly astronomer could have seen the Crab Nebula (even if telescopes had been invented) because there was no such object in the sky. But for a few weeks during the summer of 1054, this shattered orb outshone any of its compatriots and would have surpassed Venus as the beacon of the predawn horizon. Unbeknownst to our medieval ancestors, a star had already exploded some 6000 years earlier, in the direction of Taurus. Because the explosion had taken place 6000 light-years away, the evidence of this star's demise did not burst forth upon earthly skies until the evening of the fourth of July 1054, whereupon the Chinese court astronomers duly took note. The Chinese had provided such detailed records that by 1921, astronomers believed they knew the location of this "guest star" and noted its proximity to the well-known nebula. That same year, it was discovered that the Crab Nebula was expanding and apparently had been doing so for nearly a millennium. It didn't take long for astronomers to put 2 and 2 together.

I was so fascinated by this rare convergence of historiography with astrophysics that I didn't notice my entrance into the nebula itself. This surprised me, because I had expected a strong jolt as I crossed from the undisturbed surroundings into the zone occupied by the outward-rushing debris. As I had learned during my very first traversal of the Milky Way, no star lives in a vacuum, and when a star is audacious enough to explode, it must push its surroundings out of the way. At typical explosive speeds (covering 1500 kilometers every second, if not more), the quiet gas that envelops the star cannot anticipate what is about to befall; the blast overruns it without warning. Like the famous horror film in which decent citizens are overcome by the lumbering swarm of zombies and then become zombies themselves, the swept-up atmosphere is shocked into sudden motion and inexorably becomes part of the blast, plowing in turn into its surroundings and infecting *them* with unstoppable motion. The signature of such a shock wave is a sharp increase in pressure—this is exactly what a sonic boom is—and, often, the radiance that accompanies flash-heated or disturbed gas. I *must* have crossed the shock, I thought; the fact that I had felt nothing perplexed me.

Not that the existence of a shock had actually been demonstrated by my observer-colleagues on Earth. The environment of the Crab Nebula had been regarded as one of its deeper mysteries. Astronomers were always looking for its "shell," and not merely for the purpose of justifying its name. Optical, radio, and X-ray observations—any of these should have given evidence of the shock—all came up empty. When the nebula's mass had been toted up in terms of the gas that one could easily "see," it fell short by at least four times the mass of the Sun. Only stars more massive than a certain threshold were supposed to explode like this, and the four (some said six) missing solar masses constituted the difference between what was seen and what the theorists expected. Some conjectured that the "shell" was not really part of the explosion but, rather, consisted of slowly moving gas that had wafted off the star's surface during the eons before it blew up, when it was not a hot star but a cool red giant. They contorted their reasoning to find ways in which this effluence could somehow shield itself from disturbance to the point where it was invisible. Others said that the shell was there, and was actually rushing out at speeds several times greater than the expansion speed of the observable nebula, but that it was too faint to see. Why this should be so was anybody's guess. To me, the most plausible explanation was that the Crab lived in very sparse surroundings, like the interior of one of those hot stellar bubbles, superbubbles, or chimneys I had seen during my trip to the Galaxy's center. Such bubbles could be blown, over millions of years, by just the sort of star that the Crab once was. Maybe the Crab-star had evacuated its own neighborhood, and now there was little for its debris to run into.

10

Crab II

I was confident that I would soon encounter the familiar nebula, but as I continued onward without finding a trace of the exploded star, I became worried. The space around me seemed eerily empty. Debris in any form—if compacted into dense enough clouds—would have been hard to detect (assuming I didn't run into it!), but not so the luminous features that were easily visible from Earth. As I searched in vain for a landmark, it dawned on me that the reason for this desolation was profound. I was not in the Crab Nebula I remembered from Earth; I was in its ruin. I was disheartened, though I should have known. Ninety thousand years of continuous expansion had sapped the nebula's vitality. The debris from the star was now well mixed with ordinary interstellar matter, and its explosive energy had spread over a volume thousands of times larger than the historical Crab Nebula.

I knew there would still be a neutron star to study, if only I could locate it. I looked around—there were hundreds of faint stellar-looking objects that could have been candidates. At this age, the neutron star should have been very faint at all wavelengths except perhaps the radio band, where it might have shown up as a slowly winking pulsar.

I was debating whether to commence a half-hearted radio search or to give up entirely when I remembered an event that had made an impression on me as *Rocinante* approached the location of the Crab. I had spotted a new stellar explosion, off to starboard, and had noted in my journal that it exhibited many features in common with the sort of explosion thought to have produced the Crab. As I moved through the Galaxy, I could tell from its changing displacement against the background of more distant stars that it was less than a thousand light-years away. The remarkable coincidence had pleased me. The chance of a second such explosion, this close in time and space to the first, was miniscule. Now I viewed it as a godsend. What excellent luck! I could be there in less than a thousand years of Galaxy time and, if it did prove to be another Crab, experience the familiar nebula as though it had been reborn. I set course immediately for the nebula I dubbed Crab II.

From 30 light-years out, Crab II produced a ghostly effect. It appeared as a vast, dimly shining panel, taking up as much space on my sky as the Big Dipper does on Earth's. Its total luminescence amounted to only one-hundredth the brightness of the full Moon, spread over an area equal to nearly 1600 full moons. Near its center I caught a glimpse of my destination, the neutron star, appearing as any ordinary star of magnitude zero, just as Arcturus, Vega, or Capella. This neutron star was young and still shone brightly.

The nebula, like the familiar old Crab, had an oblong shape that appeared more articulated the closer I approached. In addition to the texture created by the luminous filaments, Crab II had a distinctive architecture. Two deep, rounded indentations cut into the nebula, giving it the appearance of having a waist. I could clearly make out that the constriction was three-dimensional, pinching the nebula all around. The filaments girdling the indentation had a very slightly "off" color, compared to the other filaments, and I likened them to whalebone stays corseting a satin-clad figure. The metaphor seemed apt. The expanding nebula was being held back, though not stopped, by the constriction, while in the perpendicular directions, where it was unconstrained, it appeared to expand freely. For the most part the filaments seemed to form a random network, but there were a few places where it was hard not to visualize a greater degree of organization. At one place in particular, filaments were arranged in such a way that they seemed to describe a tubular conduit. Nothing seemed to be flowing through it, yet it was hard not to attach a dynamical significance to this sharply outlined decoration, if only as a symbol of what I now saw was a delicately squeezed and shaped explosion.

I was not long to have the luxury of such a global view. Almost without warning, I found myself immersed in the sea of glowing filaments. As I approached the nebula, what had struck me even more than the shape had been the colors. Now I was overwhelmed by the iridescence of the scene. Filaments shone with the familiar rich green of oxygen and the reds of hydrogen, sulfur, and nitrogen, but many more subtle hues could also be discerned. Of course, each time an electron popped from any one orbit in an atom to a lower one, it emitted a very distinctive color, and there were many, many orbits in each type of atom. Also, many of the atoms had lost one or more of their electrons, and each of these needy atoms—ions—had its own assortment of orbits. Thus the array of colors was staggering, and I also noted the ultraviolets of hydrogen, carbon, and helium, the yellow of helium, and the violets of oxygen and neon. The individual colors were all familiar, the stuff of spectroscopy class in grad school. But something seemed odd: The mix was different from what I had come to expect.

The peculiar combination of hues and their relative intensities—let's call it the "spectrum" of "lines," now that we are going to do something quantitative with it—depends, more or less, on two factors. One is how well or roughly the atoms are treated. The other is the mixture of different chemical elements, or "composition." The atoms in Crab II's filaments were being disturbed in a couple of ways. Most important, they were being tickled by that bluish glow that I had remarked on earlier. Atoms see light as chopped up into its constituent particles, or photons, which constantly move around at the speed of light (naturally) and sometimes hit electrons. When that happens, the photon is absorbed and can either knock the electron into a higher orbit, whence it drops back down and emits new photons with very specific colors, or knock the electron clear out of the atom. In the latter case, the precise hues are emitted when the freed electron finds an atom to attach itself to and drops down through a sequence of orbits as it heads for home. The photons that make up blue light do not pack enough punch to jostle most of the electrons out of their preset orbits; as a result, they do little. But it took more than just blue light to give the nebula its garish cast. My eyes settled on the blues because they filtered this light for its visible content, but the true spectrum was a continuum of colors from the radio (which had even less effect on electrons than the blue photons did), through the infrared and all visible colors, and on past the violet into the ultraviolet, X-rays and gamma rays. Tipped as the spectrum was toward the more energetic rays—on beyond violet—there were plenty of photons capable of ionizing and otherwise disturbing the atoms of the filaments. And as though that weren't enough, there was a second mechanism that also seemed to be operating: The atoms suffered collisions with freely moving electrons, or even with other atoms, that also knocked orbiting electrons out of their appointed rounds. Some of this activity came from the chaotic motion of heat, acquired either from the very fast electrons that (I was soon to learn) produced the bluish glow or as the filaments were warmed by basking in the blue radiance itself. The rest of the motion had its origin in the explosive energy with which the filaments had been shaped and expelled from their point of common origin.

Considering the harsh environment, the atoms inside the filaments of this reincarnated Crab were not treated too badly. The fact that they were not bashed to pieces and completely dismantled allowed them to produce the rich spectrum of hues that I experienced and enabled me to make a detailed study of their properties. By analyzing, comparing, and keeping track of the strengths of all these spectral lines, I could deduce both the nature of the filaments' excitation and their chemical composition. Here I had a surprise. The reason why the colors made up such a strange mix was that the filaments were overwhelmingly composed of helium, with small admixtures of oxygen, carbon, and everything else. Hydrogen—normally the dominant species—was only 10 percent by weight in the filaments, and it was even rarer in the gas that composed the "corset stays" binding the nebula's waist. That explained why their colors were even stranger than those of the "normal" filaments.

I had never encountered gas with such a weird composition before. Helium is a rare element on Earth, but that's because it doesn't react with anything and most of it evaporated into space during Earth's formation. What little helium can be scavenged on Earth has been produced as a by-product of radioactive decay and trapped underground in pockets of rock. It is more common elsewhere in the Universe. The element was first discovered in the spectrum of the Sun (hence the name, which is derived from *helios,* the Greek word for "Sun"), It makes up about 27 percent of the sun by weight, with nearly all the rest consisting of hydrogen. These ratios were what I was used to. One found them nearly everywhere—except on Earth, where much of the hydrogen had escaped as well. The near universality of the helium-to-hydrogen ratio is easy to understand. It bespeaks the manner in which most of the helium in the Universe had formed: in the crucible of the Big Bang, just a few minutes after time-zero. There seemed to be only one way for this makeup to have been distorted toward the extreme of nearly pure helium: via nuclear reactions inside the star that had exploded. As it turned out, my first encounter with helium-rich gas proved to be only the tip of an iceberg. In other settings, later, I was to encounter chemical compositions far more bizarre that held even more important clues to the great cycle of matter, as mediated by the deep interiors of stars.

11

Strange Light

I was curious about the nature of the sickly bluish glow that seemed to be everywhere. At first it had reminded me of the glow that sometimes envelops one when one is walking down a street in a thin mist, except that the glare of a mist is all second-hand light, scattered from some other source. In this case there was no light from a nearby street lamp to scatter. This vapor was intrinsically luminescent. The color (as I perceived it) and the sensation of being surrounded by an irradiant medium reminded me of snorkeling in a South Sea lagoon rich with clouds of bioluminescent creatures—they put out a very similar kind of glow, though more blue-green in tint. The medium also had that mottled look one finds in a bioluminescent sea, with striations and bright patches where (perhaps) the medium had been disturbed. Only this wasn't bioluminescence. As I sampled the medium that occupied the spaces between the filaments, I saw that its active ingredient, or at least one of them, was an extremely hot gas of electrons.

The glow, however, was not at all the radiation of a typical hot gas. No matter how I plotted and parsed its spectrum, I could not get any indication of temperature. It seemed to have all temperatures and no temperature at once. I remembered that I had seen a similar glow near the Galactic Center's black hole. I measured the speeds of the electrons and found that they were charging around at so close to the speed of light that at first I had trouble telling them from photons. Rather than traveling in straight lines, though, they were executing tight gyrations. I knew immediately that the agent of this motion had to be a magnetic field and that the glow had to be what my colleagues called synchrotron radiation.

A quaint name, "synchrotron radiation." It referred originally to an antique type of atom smasher. When an electrically charged particle moves through a magnetic field, it is deflected from its straight-line path onto a circular path or helix. Early particle physicists found this circular motion convenient, because it provided a way of keeping fast-moving particles confined to their experimental apparatus and stopped the particles from crashing through the walls and doors of their labs. In the device called a synchrotron, an alternating electrical impulse was synchronized with the circular gyrations in such a way that the electrons were gradually accelerated to speeds approaching the speed of light. But the experimenters began to encounter problems. A gyrating particle emits light. And when the electrons got really close to light speed, the physicists found that most of the energy they were pumping in through electric fields was coming straight out again, in the form of a glow that they called synchrotron radiation.

Thus synchrotron radiation was originally an investigator's nuisance. But it was a key to understanding Crab II, because when I measured the totality of the bluish light, I found that it, not the sharp hues of the spectral lines, provided the dominant illumination of this nebula. I also saw that the glow got stronger and harsher, the closer I moved toward the neutron star. Now I was weaving my way through the thicket of filaments, trying to avoid colliding with the denser gas. Essentially all of the matter was concentrated in the filaments. This was debris from the exploded star, and it had to have enough bulk to account for several solar masses of matter. In contrast, the glowing medium between the filaments was a far better vacuum than any region of interstellar space I had sampled so far. I wondered why this super-hot matter didn't just dodge between and around the filaments and escape into the surrounding space, which, after all, seemed to have little means of hemming it in. What I saw instead were bubbles of the hot fluid bulging up against the filaments or wrapping around them, but then being held back as though by some elastic membrane. The strange, springy behavior of magnetic fields came to mind, as it had when I visited Cygnus X-1 and tried to fathom how the swirling gas managed to lower itself toward the black hole. I knew the magnetic field had to be there, because it was a necessary ingredient for the generation of synchrotron radiation. I also knew how to make its structure visible. Through polarized lenses I could map out the lines of force. The magnetic field here was indeed the agent that turned this luminous near-vacuum into a kind of jelly, its springy, membranous quality permitting the network of filaments to act as a cage with few bars.

At less than a light-year from the neutron star, I had left the cage of filaments behind. It was all non-thermal glow here, and the intensity was becoming almost unbearable. *Rocinante* was being bombarded increasingly by X-rays and gamma rays, which indicated that I was nearing the source of the super-fast electrons. I needed no more confirmation that the ultimate energy source of this glare was the neutron star itself, but I got more confirmation nonetheless. I saw now that this luminous medium was really a wind, emanating from the neutron star and spreading out predominantly along a plane that—I was later to learn—marked the equator of the neutron star's rotation. Despite the intensely energetic nature of this environment, I saw strangely beautiful and delicate structures. In the exact central plane of the wind there was a very thin sheet that was especially luminous. The sheet was rippling like the surface of a pond, concentric swells coming outward at me one after another. These ripples were traveling at about a third the speed of light, but the distortions of space and time predicted by Einstein's theory made them seem to approach even faster and produce a whiplash effect as they passed. Well above this sheet, in a direction that I surmise lay above the neutron star's polar axis, were sprightly little flares, like St. Elmo's fires at the top of a ship's mast, dancing amid jets of nearly invisible plasma.

Just a couple of tenths of a light-year from the neutron star, I crossed some sort of boundary and the character of my environment changed completely. I was no longer immersed in the glowing medium, yet I could see it everywhere I looked. I was apparently inside a bubble with a luminous wall. The wind that now surrounded me was no longer radiant, but it was incredibly fast. Its speed was within one part in a trillion of the speed of light, and I saw that it also carried a magnetic field. But the electrons sweeping past me were not emitting much synchrotron radiation, because they were hardly gyrating in this field. Instead they seemed to be moving along with it. It was only after they plowed into the luminous wall—which I now saw was a shock wave—that their orderly motion turned to chaos and the garish light streamed out.

The wind was powerful. It carried 100,000 times the power of the Sun, all of it coming from what still appeared to me as a bright point of light in the distance. This neutron star somehow powered the luminosity of the entire nebula—wind, blue glow, filaments, everything. But how?

From this distance the neutron star appeared about as bright as the full Moon, which meant that it was shining with the luminosity of several hundred suns. It was emitting much less light than it was pumping out in the power of its wind. But if it were only 20 kilometers across, it would have to be very hot all the same: 6 million degrees, if the light leaked out from its interior like a normal star. I searched for the confirming signature, the X-ray "color" that goes with such a temperature, but found that the spectrum of colors was bizarre. In addition to X-rays, I saw far too much ordinary visible light. The radio waves coming from this thing were even more remarkable; if I had measured them alone I would have concluded that the temperature exceeded a trillion trillion degrees!

The pattern of light coming off the neutron star was also very weird. Unlike that from an ordinary star, the light wasn't coming off evenly in all directions. As I shifted my position away from the plane of the wind, the brightness increased, then decreased, then increased again. Somehow the star's light was concentrated into parallel bands that circled the globe like bands of latitude. I recalled that when they were first discovered, neutron stars had briefly been known as LGMs for "little green men," and for a moment I imagined that these bands of light could be artificial. Were they perhaps navigational aids? Then I remembered the real reason for the early wild speculations about these objects: the pulses. Of course! The neutron star of Crab II was a pulsar. It was a relatively fast one, spinning out two oppositely directed, hollow beams of light 30 times a second. When I hadn't looked too carefully, the bands of light had appeared steady. But with foreknowledge and some imagination, I could tell that I was being strobed. I headed in for a closer look.

This time I knew my limitations in advance. Given that this neutron star's mass was about 1 ½ times that of the Sun (if neutron stars whose masses had been measured—those in binaries—could be relied on as a guide), I knew that I could get slightly closer than I could to Cygnus X-1 without turning my insides out. (Whether I could have ventured closer to SS 433 I will never know, because cowardice had gotten the better of me there.) Accordingly, I parked 6000 kilometers away from the pulsar, carefully positioning myself out of the blinding glare of its gyrating beams. From this distance the neutron star was half the size of the full Moon as viewed from Earth and could be comfortably observed with binoculars or my small telescope.

I had half expected to see a mirror. Because of the enormous gravity, neutron star surfaces had to be exceedingly smooth. Any "mountain" higher than a few millimeters above the surrounding plain would either melt and drizzle away or crack through the crust beneath it and sink into the mantle. Scaled up a thousand times to Earth size, this would be equivalent to there being no peaks higher than 10 meters above sea level: The whole Earth would be topographically flatter than Florida. Also, because of the extreme densities—thousands of tons per cubic centimeter right up near the surface and rapidly increasing with depth—the matter in the crust of a neutron star would exist in an odd state. The electrons would be quite fluid with respect to the atomic nuclei, never quite being able to decide to which one they belonged. This could be a recipe for a metallic surface, or at least a good electrical conductor, or perhaps a crystalline substance—I could never quite work it out from the treatises I had read. I had visions of seeing the background stars, or the pink and green glowing filaments of the nebula, or (in my more whimsical moments) my own face reflected in a shiny globe, distorted to bizarre proportions by its convex shape and the bending of light rays in its intense gravitational field.

Thus it was a bit of a letdown to discover that any visible characteristics of the neutron star's surface were washed out by its heat radiation. The ball glowed brightly in X-rays, though with an output not nearly so great as that of the pulsed radiation. I estimated the temperature of the glowing sphere to be a few hundred thousand degrees. I knew that this was not the radiant energy from nuclear reactions; the star was dead from that point of view. It must have been the residual heat from its cataclysmic birth, leaking out from its interior. This made sense—the neutron star had existed for less than two thousand years.

I could see from here that the pulsing radiance had little to do with the stellar surface. The fireworks were all occurring much farther out, most of the action being concentrated about 1600 kilometers away from the neutron star. At 160 times the radius of the star itself, this was really quite far away. If one scaled up the neutron star to Earth size, the activity would be located at three times the distance of the Moon. At the dizzying rate of the pulsar's spin, the active zones blurred into shimmering bands of light, but with a series of snapshots I attempted to freeze the action and see what was going on. The site of activity was still indistinct, but most of the luminosity appeared to come from a pair of oval rings that lay above two bright spots on opposite sides of the neutron star and rotated with the star as though tethered to it. The rings were not steady and well defined; they danced, flickered, and changed shape, sometimes brightening and at other times nearly fading from view. They reminded me of the aurora borealis, not the ground-based impression one gets of vast luminous curtains of pink, blue, and yellow being ruffled by some heavenly wind, but the view I once had from above, from a satellite orbiting the Earth. Fast particles—cosmic rays—streaming down from the Sun into the upper atmosphere, were channeled along magnetic lines of force and converged in a bull's-eye encircling the Earth's magnetic north pole. The aurora formed where the particles hit and disturbed atoms in the upper atmosphere.

This analogy was obviously imperfect. With difficulty I could trace flares and streamers of light connecting the rings to the spots on the surface of the neutron star, which were a little bit hotter than their surroundings. Frequent discharges—sparks—erupted. In the pulsar the particles seemed to be coming from near the surface, or perhaps they were being created spontaneously by intense electrical disturbances in the space between the star and the luminous rings. And these particles were being flung outward, toward and through the rings, not streaming in from space as in the aurora. In one respect the two systems were similar, though. The pulsar emission was coming from near the neutron star's magnetic north and south poles.

Where did the magnetic field come from and why was the pulsar spinning so fast? The second part of this question was the easier one to answer. This object was so dense, so compact, that it must have formed through the collapse of a much larger object. I knew the theoretical story, how the core of a massive star must have lost its resistance to gravity and fallen in on itself. But in this case the details hardly mattered. The Sun rotates, every star rotates, and if a rotating object shrinks, it must rotate faster. Conservation of angular momentum. The figure skater drawing in her arms to spin faster. (How many times had I heard *that* analogy, but was there a better one?) For the magnetic field, the process was more complicated. Stars also have magnetic fields and are good electrical conductors. If a star shrank and the magnetic field failed to keep pace by becoming stronger, then one could calculate that an electrical force would develop and grow. This would drive electric currents, and the currents would strengthen the magnetic field. So the magnetic field would have to grow, whether the star liked it or not.

The analogy that had served me well in my efforts to understand the disk in Cygnus X-1 also worked here. Magnetic lines of force in stellar plasma were like colored stripes in salt-water taffy. They stretched if the star stretched. They twisted if the star twisted. If the star shrank, the stripes would also shrink and cram closer together—*voilà*, a stronger magnetic field. But that may not be the whole story. I hadn't actually witnessed the collapse that created the Crab II neutron star. What if it hadn't been perfectly symmetrical? What if pockets of trapped heat had forced their way to the surface in the form of bubbles and rising blobs? Then, for a brief time before it settled down, the interior of the collapsed star could have been a churning mess of stretching and swirling fluid. The taffy analogy applied once more: The field could have strengthened in this way, too.

Whatever events had led up to the present condition of this neutron star, its extraordinary attributes were undeniable. Its magnetic field was enormous. A trillion times larger than the terrestrial magnetic field that sufficed to channel the particles of the aurora borealis and align the compasses of seafarers. Billions of times larger than the magnets that used to hold notes to my refrigerator back on Earth. And this magnetic field—a gigantic bar magnet embedded in a 20-kilometer-wide bowling ball—was being spun around at 30 times a second. Out 1600 kilometers from the neutron star, the lines of force had to swing around so fast to cover the circumferential distance that their speed approached the speed of light. At this point the difference between magnetic and electrical forces became blurred, fast particles no longer remained under control in the neutron star's sphere of influence, and—not to put too fine a point on it—all hell broke loose. Particles, electromagnetic energy, and radiation streamed forth, the first two ultimately destined to power the bluish glow, the filaments—the entire nebula.

Still, I wasn't completely satisfied. What was the ultimate source of this power? Unlike Cygnus X-1, it couldn't be the strong gravity of the neutron star, at least not directly. No matter was spiraling toward the star, a requirement if gravity were to be relied on to generate energy. The magnetic field, strong as it was, wasn't strong enough to power this 100,000-fold solar luminosity for 2000 years. It was at best only an enabler, a conduit through which some much more abundant active agent flowed. Only one viable power supply remained: the energy stored in the neutron star's spin. This was exactly the deduction that the Cornell astronomer Tommy Gold had made in 1968, establishing beyond any reasonable doubt that pulsars were spinning neutron stars. I was able to repeat his experiment on the spot. If the energy were stored as rotation, then as the pulsar used up this energy, expelling it into space, it should spin more and more slowly. And this is exactly what it did. Every year, its spin period increased by 1 part in 2500, so that in a couple of thousand years it would be spinning only 10 or 20 times a second. As it dragged its magnetic field around less and less violently, its expulsion of energy should slow down as well. No wonder the aged Crab Nebula that I had first visited, though bigger, was a dimmer, drabber place.

Thus the Crab II pulsar, and the Crab Nebula's pulsar before it, were huge and powerful flywheels. But a flywheel is just a repository for energy. Some motor must have spun it up. In this case, that motor was no mystery. It was the gravitational pull that had collapsed the star in the first place. The accelerating spin of the skater comes ultimately from the pull of her muscles, as she draws her arms inward. It takes calories to spin up. The spin of the star comes from the pull of gravity. It takes contraction, inflow, or accretion to spin up. I was back to gravity again. Maybe it was the key to everything, after all.

I was nearing exhaustion. I had had no idea that the environment of a neutron star could be so complicated. Working my way inward, I had encountered layer after layer of astonishing phenomena, each one more exotic than the last. The pulsar was in some ways the most extraordinary thing I had seen on any of my excursions. Nothing was quite so bizarre as the horizon of a black hole. But in the case of the black hole, I could at least fathom a connection between the supply of matter and the output of energy, even when its details were obscured, as they had been in SS 433. By contrast, the pulsar was so "clean" that it seemed almost magical. It was nothing like a furnace; no fuel was being consumed to make it shine. Its spin, acquired long ago from energy extracted by gravity and set aside during its sudden collapse, empowered it to shine as a lighthouse beacon for thousands of years. This energy supply was turned into a scorching wind and harsh radiance through mysterious force fields that seemed to act over great distances, their pulleys and levers masterfully concealed from view.

12

The Ends of Equilibrium

My universe seemed to have grown another notch more complex, and still I hadn't accomplished my mission. Witnessing a pulsar was not the ultimate objective of my quest here. Gravity hadn't triumphed utterly in the Crab II pulsar, or in the old Crab pulsar, for that matter; there would be a phase beyond the spinning flywheel. Once the pulses slowed down, after the rotational energy had been spent, even if some day the magnetic field died away, the neutron star would still be there. It was a body that had imploded to within a few kilometers of sinking beneath its own gravitational horizon and becoming a black hole but had stopped before it had gone too far. A body that thumbed its nose at gravity despite having no internal source of nuclear or other power to keep it hot. Yet here it was, as stable as a rock, and certain to remain so indefinitely.

I needed to find the origins of a neutron star's equilibrium. That was why I had come to the Crab Nebula in the first place, although so much had happened since my arrival that I had almost lost sight of this goal. I wasn't even sure what I meant by equilibrium. An equilibrium involves stasis, persistence, something approaching permanence, but is it necessarily static? Did the well-behaved, nearly circular orbits of stars around the Galaxy's center—or, for that matter, the orbits of planets around the Sun—constitute equilibria? I supposed so, in the sense that there was a balance of forces—gravity against centrifugal—or, more precisely, a simple and steady kind of motion, an acceleration, that resisted gravity's efforts to draw each star toward the Galaxy's center of mass. But in another sense, the answer was not so clear. There could be a structural equilibrium that was static: The shape of the Milky Way's disk, for example, sketched in by the trajectories of *all* the stars that shared the generic similarity of executing circular orbits in the same flat expanse. The motion of an individual star on an orbit was not an equilibrium of this kind, because the star's position around the circle changed with time.

Clearly, the neutron star must be an equilibrium of the second kind. I was not interested so much in the individual motions (or lack thereof) of the particles that made up the neutron star. But, viewed as continuous matter, whatever substance made up the neutron star kept its shape and resisted gravity with remarkable rigidity.

Rigidity—or stability—was another aspect of equilibrium I had to worry about. No equilibrium was worth its salt if it could be overthrown by a sneeze. A playing card standing precisely on edge and a sharpened pencil balanced on its point are both equilibria, surely, but not terribly useful ones. There needed to be some feedback that would oppose any *modest* attempt to upset the balance. I don't think I was asking too much. I would grant that if the Milky Way collided with another large galaxy, its well-ordered stellar disk might not survive the disruption. Not that the stars in the two galaxies would collide physically—I'm certain they wouldn't. The disruption would be more subtle than that. Stars originally belonging to one galaxy would be tempted away by the gravitational lure of the other. Forces changing rapidly in strength and direction, as the remnants of the two galaxies jockeyed for position, would defeat orderly motion, throwing it into disarray. On the other hand, the Milky Way's disk had better be robust enough to weather a minor disturbance: the intrusion of my spacecraft, for example; or an errant star flying in from intergalactic space; or even the impact of a moderately massive black hole; or a globular cluster containing a million stars, crossing the disk. The latter might create ripples, modifying the disk slightly, but should leave the basic structure intact.

I pondered the feedback that might lend a disk of stars the kind of stability I was looking for. It would have to be the conservation of angular momentum: the same effect that had spun up the pulsar. If the disk expanded in radius, its stars would no longer be orbiting fast enough to withstand gravity, and the disk would sink back toward its original size. If it were compressed, the stars would speed up in their orbits and the disk would spring back. But then I remembered that a disk of stars, in isolation and orbiting under its own gravity, is *not* stable. I had been through all of this reasoning before, when I had traversed the Galaxy and wondered about the spiral arms. There were other ways to perturb a disk than simply to expand or contract it, and some of these alterations would lead to unruly outcomes. Groups of stars would take sides, gang up on each other, trade angular momentum back and forth, and in so doing disrupt the orderly structure. The disk could slosh, bend, or break. The truly stable entity was the disk plus halo—with emphasis on the halo, which had to contain most of the mass. A halo would act as a moderator, preventing things from getting out of hand.

Like the disk, the halo consisted of moving stars, but they were not marching lockstep in circular orbits; they were moving randomly, every which way. The halo was also a kind of structural equilibrium, and thinking about it gave me a better opportunity to visualize how feedback might lead to rigidity. I imagined that I could somehow take the Milky Way's halo of stars and stuff it inside the toe of a thick wool sock. What resistance would I discern as the stars bumped into the wall of the sock? There would be so many stars hitting the wall and bouncing off that it would feel like the steady pressure against the wall of a balloon. If I grabbed the open end of the sock and squeezed the trapped stars into a smaller volume, the halo would push back. The pressure against the inside of the compressed sock would be higher, because stars would be hitting the sock more frequently in the confined space *and* because each star would be moving faster. The increase in speed, I realized, followed from a generalized version of the conservation of angular momentum. If I allowed the sock to expand, the opposite would happen: Stars would be moving more slowly and scoring fewer impacts, generating less pressure. So far, this behavior was just like that of any gas—the air inside a balloon, for example. But for the stars of the halo, there was an additional complication. As the same number of stars was confined to a smaller volume, the gravitational attraction inside that volume would increase. The big question was whether the increase in the stars' speed, which made them less susceptible to gravitational collapse, would outstrip the increase in gravitational attraction, which made them more susceptible. There was nothing to do but to put on my theorist's hat and do the calculation. This I did, and the results were clear. Gravity did not prevail against the feedback of increased motion. It was angular momentum again that saved the day.

These thought experiments were getting complicated and threatened to take me too far afield of my principal concern. I had to focus. What was it that opposed the tremendous gravity inside a neutron star? It might be motion of some kind. The chaotic motion of its atoms holds up a star like the Sun, much as the chaotic motion of its stars holds up the halo of a galaxy. In the short term, the feedback in the Sun is exactly the same as that in the Milky Way. After all, the Sun really is made out of gas, very similar to the air in a balloon. But a second level of feedback is needed in the Sun's case. Unlike a system of stars, the atoms in the Sun are always colliding and losing energy, and this energy leaks out in the form of the Sun's radiance. The losses are constantly replaced by nuclear reactions going on in the Sun's center, which keep the Sun shining steadily. If these reactions faltered, the Sun would begin to deflate under its own gravity, but then—like the stars trying to conserve their angular momenta—

the atoms would speed up. The nuclear reactions, being very sensitive to the speeds with which atoms slam together, would immediately pick up their pace, providing the feedback to keep the Sun stable.

The solar analogy was appealing, but it proved to be a dead end. I knew that nuclear reactions don't go on inside neutron stars. Neutron stars are burnt-out entities, *sans* thermonuclear fuel. Perhaps this meant that they had to be able to survive without relying on the motions of their constituent particles. Was that untenable? After all, motion is not the only thing capable of counterbalancing gravity. The universal quality of gravity, in every single one of its manifestations, is simply that of attraction. So any form of repulsion would suffice, if it were strong enough and provided the right kind of feedback.

The Earth resists collapse not because its particles are in lively motion but because it is made of very resilient material. Gravity squeezes the layers of solid and molten rock inside the Earth, but they are so resistant to compression that they can support the weight handily. Even outside the high-pressure environment of a planetary interior, the atoms in these materials—the electron clouds enveloping their tiny nuclei—nearly touch. Any additional squeezing jams the electron clouds together and even makes them interpenetrate slightly. Atoms fiercely resist such overlap. Each atomic nucleus, an infinitesimal BB made of protons and neutrons (only one ten-thousandth the size of its electron cloud!), is electrically charged. Stripped of their electron clouds, they would repel one another with a vehemence so great it is hard to imagine. Compared to their mutual gravitational attraction, the full-blown electrical repulsion would be 10 trillion trillion trillion times larger. But the atoms in ordinary matter do not fly apart violently, because the electron clouds neutralize this repulsion. They possess electrical charges that are exactly equal and opposite to the charges of their nuclei. Thus, in their relaxed states, neighboring atoms coexist benignly, even enjoying a slight degree of attraction brought on by mutually induced distortions in the electron clouds. When atoms are forced together under such tremendous pressure that the electron clouds begin to overlap, however, some of the perfect screening is sacrificed. A little bit of interpenetration is all it takes for the repulsive force to reappear in just the right amount to resist the pressure.

Thus electrical repulsion is the force that opposes gravity in the Earth and other rocky planets. The feedback in the Earth's equilibrium comes from the fact that the degree of interpenetration—and hence the amount of repulsion—can adjust very precisely to balance the squeezing force of gravity.

But this clearly wouldn't work in a neutron star. In the Earth, the electron clouds overlap slightly. But if the pressure were only 100 times higher than it is near the center of the Earth, the electron clouds would overlap so thoroughly that they would effectively coincide. This is a quantum mechanical no-no called *degeneracy*. If you try to force quiescent electrons to occupy the same space, they will slip out from under your fingers, scurrying away with alacrity to avoid such embarrassment. The effect is remarkable, and it means that motion appears once again as the dominant counter to gravity.

This kind of motion has a very special property. Unlike the motion inside a low-pressure star like the Sun—the motion of heat—the motion associated with degeneracy does not lead to the loss of energy through radiation. Therefore, no nuclear reactions, or any other sources of energy, are necessary to maintain the equilibrium. The speeds simply increase as the matter is squeezed—and that feedback, alone, is sufficient.

It seemed fitting that a neutron star, with all of its other strange properties, should be supported by such a very strange form of pressure. It isn't the electrons' degeneracy motion that supports the neutron stars in the Crab and Crab II, however. The pressure needed to support the weight of a neutron star is a trillion trillion (1 followed by 24 zeroes) times higher than the pressure at the center of the Earth. The atoms are squeezed 100,000 times closer together than in ordinary matter. At these densities there is no room for electron clouds at all. The atomic nuclei themselves overlap, the electrons having long since been absorbed into the protons to compose neutrons—hence the popular name of these objects. Neutrons also have the property of resisting degeneracy, and it is their quantum skittishness that provides the main counter to gravity. A more prosaic form of repulsion enhances the motion induced by degeneracy. This is analogous to the electrical repulsion of ordinary matter under pressure, except that this time the force is not electrical but the even stronger nuclear force that binds the chemical elements. A neutron star is a little like the nucleus of a humongous atom, with an atomic weight of 10 followed by 56 zeroes. Compare that to normal matter: Carbon's atomic weight is 12, that of uranium, the heaviest naturally occurring element on Earth, 235. The nuclear force is apparently claustrophobic; when too many nucleons (protons or neutrons) crowd together, it switches from attractive to repulsive. Neutron stars are unique in being the only atomic nuclei held together by gravity.

This visit had given me a lot to think about, but I still felt a nagging frustration. I hadn't been able to plumb the depths of a neutron star in person, so I had had to plumb them in thought. I was not pleased at being forced to adopt such a theoretical perspective when I was so close to my object of consideration and had gone to so much trouble to get there. How much had I really gained by making this long journey? My practical side argued that there was no point in worrying about that now. Equilibrium was such an important aspect of the cosmic constitution that my efforts to understand it really were necessary, whether they took place in comfort on Earth or bucking the harsh, arid wind of a pulsar.

Then I gazed back out through the nebula. In the distance the soft glow of the filaments was dimly visible behind the stark blue synchrotron glare. I could almost sense the cool debris rushing away from me in all directions. I imagined that this was the original Crab Nebula, in its heyday as observed from Earth, not the reincarnation that I had fortuitously discovered at just the right moment. I could appreciate the incredible energy that had been released near this point—first, all of a sudden in the explosion witnessed in 1054, and then, over the thousand years since, supplemented by the steady braking of the pulsar's spin. It was not the most beautiful of nebulae. But my antipathy toward the Crab Nebula had vanished, to be replaced by respect and a measure of awe. I had explored the concept of equilibrium almost as an afterthought, because there had been so much else to see. The relationship of the familiar-looking filaments to the macabre "bluish glow"; the sensation of crossing through the shock into the inhospitable inner precincts of the rushing wind; the appearance of the pulsar itself. Despite all I had read before about these exotic bodies, no description had prepared me for their alien character.

I knew that no cosmic object took cognizance of human presence. (I suspected that Earth had only tolerated us and was probably regretting the decision.) But this pulsar, possibly because it had carved for itself such an austere and harsh environment, seemed more indifferent than the rest. Even more indifferent than the black holes of Cygnus X-1 and (perhaps) SS 433, with their warm binary companionship and their sloppy disks of opaque, swirling gas. By comparison to these dog-like attributes, the Crab II pulsar was definitely feline. And in the end I had found this virtually indestructible object—a neutron star—that had emerged out of the violent death of a great star. Maybe it was my own perversity that had led me to explore equilibrium at its extreme outpost, in a body just this side of being a black hole. After all, I could simply have visited the Sun for an example of equilibrium. But seeing equilibrium existing so close to the edge brought home to me how crucial a concept it was. Gravity was a flexible force. It could generate liberating motion and even shoot jets out into space, or it could clench its teeth and fight the resisting pressure to an immovable draw. It all depended on whether there was the right kind of feedback.

Seeing this endpoint of stellar life also roused in me an understanding that it was not sufficient merely to study how things are in the Universe; we must also investigate how they became what they are. Before there was a Crab Nebula or a Crab II

Nebula with its pulsar, there was an ordinary star that shone, much like the Sun (if considerably more brightly), for at least a few million years and was held up by the heat of its internal gas. Where did that earlier equilibrated globe come from? How did gravity pull its raw materials together in such a way that it staved off final collapse at least for a while? I decided on the spot that I would seek out the opposite end of a stellar life, to find out how material is assembled into a star in the first place. Could I help it if the best place to discover this also turned out to be one of the most charming and picturesque locales in the Milky Way? After all the theorizing I had just been forced to do, I deserved a vacation.

Part Four

Birth

13

Orion

Orion! As a child growing up in Montana, I had dreamed about going to Orion. It symbolized warmth, growth, and a nurturing environment to me, even before I knew why it should. This association has always seemed paradoxical, because Orion is quintessentially a winter sight at my home latitudes. Every autumn, I was torn between sadness at the loss of summer heat and excitement in welcoming the crystalline appearance of the constellation of Orion, complete with its opalescent centerpiece—the Nebula. It was only on those rare mid-August nights, when I was allowed to stay up well past midnight to watch the Perseid meteors, that I had my cake and ate it too: summer's warmth *and* Orion.

There had always seemed to be an effortless grandeur to the Orion Nebula. It was not just its presence smack in the middle of what was clearly the most sparkling and easily recognizable constellation, although this certainly helped. Orion's frame of brilliant stars—Betelgeuse at one corner, Rigel diagonally opposite, the other two corners known only by Greek letters but deserving better—set the nebula apart as a work of art. And this frame of stars in turn had a frame of glowing gas, in the form of deep red arcs that stretched a good 15 degrees over the December midnight sky: Barnard's Loop, 30 Moons across. Richly colored photographs had burned its image into my imagination as indelibly as though I had seen it myself. My repeated efforts to detect its subliminal fluorescent glow were a fruitless project that had made its image even more vivid.

No such struggle was needed to see the nebula, though. With binoculars or my small telescope, the nebula filled the field of view and revealed the whole assembly to be. . . well, I can think of no better way to describe it than as a kind of matriochka, a doll within a doll. For there amid the glowing clouds was a tiny pattern of four bright jewels, the cluster of stars known as the Trapezium, slightly lopsided, not so rectangular, but unmistakable in reiterating the stellar frame of the great constellation in miniature. As a child I had found this mysterious and had naively wondered whether, upon visiting the Trapezium, I might find a tiny model Orion Nebula hidden within this frame, and so forth.

Now the thought of visiting Orion seemed less of a scavenger hunt and more like a trip to see the neighbors. It had not escaped my attention that in each successive episode of my travels so far, I was venturing closer and closer to home. I found this reassuring, not just at the visual or visceral level, but also when I thought of what Orion really was. I knew that the Solar System was believed to lie near the inner edge of one of the Milky Way's spiral arms—not a major one, but a middling spur named, in fact, after Orion. These arms, as I had understood from my earlier travels, were self-perpetuating bottlenecks in the otherwise smooth flow of stars and gas clouds around the disk. They were the urban areas of this otherwise suburban Galaxy, the places where gas and stars mixed it up and things happened. All around the Sun, one could see evidence that new stars occasionally formed, mainly, it seems, through the jostling and bumping of molecular clouds that occurs in these Galactic traffic jams. The Sun had long since left its nursery, having survived nearly 5 billion years already, and was merely passing through the Orion arm. But some nearby stars were so bright that they were doomed to burn themselves out quickly and therefore could not have existed for longer than an instant by the standards of a galaxy's life span. They had been born in the neighborhood. With some imagination, one could pick out an ill-defined belt of them surrounding the Sun at distances of 500 to 1500 light-years, as the astronomer Benjamin Gould first did in 1880. The Orion Nebula seemed to be the "buckle" of Gould's Belt, as the most spectacular of these nearby star-forming regions. I began to feel that Orion was part of the Sun's lifeline to the bustle of the Galaxy and that it could hold important lessons for me. Thus I rationalized, as adults often do, my innate desire to see the place for its own sake.

14

By the Back Door

Orion's exact arrangement of stars and luminous mists, as viewed from Earth, is so elegant that I would like to have brought my craft back toward the Solar vicinity so that I could approach the nebula from the familiar direction. But this would have been a considerable detour, because Orion lay in the same general direction as the Crab Nebula (and Crab II), though merely 1500 light-years from Earth, compared to 6000. I was already near the Crab, so I reconciled myself to approaching Orion from "behind."

This was slightly disconcerting at first, because I could not get my bearings. Orion's setting was not nearly so dramatic from this direction; in fact, it was virtually invisible. Rather, I should say that it was visible in the breach, by virtue of what it prevented me from seeing. From this direction, the Orion neighborhood presented a visual blockade, an array of impenetrable dust clouds that blotted out all view of the stars beyond. There was no bright nebula at all. The pattern of dark blotches on the sky reminded me of the rifts or "coalsacks" that had laced, split, and obscured my view of the Milky Way in the directions of Sagittarius and Scutum, frustrating my efforts to observe the Galaxy's central regions from Earth. I recognized these ragged shadows as the outlines of giant molecular clouds and tuned my radio receivers to the frequencies of gyrating and jostling carbon monoxide molecules. An image of the clouds in glowing molecules lit up my display screen, and the nebular environment immediately seemed to reacquire shape and substance. Where there was carbon monoxide gas, there were also bound to be molecules of hydrogen—pairs of H atoms stuck together—and in this case lots of them. I estimated the complex of clouds to contain at least several hundred thousand solar masses of molecular gas, mixed with thousands of solar masses of dust. This was enough dirt to mold 10 billion Earths, although I knew that little of it would actually find its way into planets any time soon.

Looking out the window once again, I traced the clouds' silhouettes by the stars that peered around their edges. For the most part, these stars were fairly distant, some even not too far from the Sun. To one side, though, a modest cluster of medium-bright stars stood out. These were nearby, and from their temperatures and luminosities I could tell that they could not be much older than 10 or 12 million years. This was commonly thought to be the first association of bright stars that had condensed out of the Orion molecular clouds; in fact, their brightest members were already gone—burnt out. The present-day fireworks of the nebula surrounded even younger stars, and they were nowhere to be seen. The Trapezium cluster, the other hot and massive stars, the glowing sheets of gas—they all had to be on the far side, the side facing the Solar System. Of course I had modern navigational aids, so I knew where I was headed. The hidden stars were there, all right, their presence manifest in the blotches of infrared luminosity—patches of heated dust—that I now brought up on my screen. There, unmistakably, was the outline of the cavity blasted out by the Trapezium, the illuminated walls of which composed the Orion Nebula itself. A second warm glow, I knew, must be the cluster of newly formed stars that went by the utterly unromantic name BN-KL, after the initials of its discoverers. This grouping was obscured to the eye both from my present direction and from Earth's point of view, and I imagined it as occupying its own cozy niche completely surrounded by the insulating and opaque molecular gas, a small cabin in dense woods.

Navigating by dead reckoning on the twin glows caused by the Trapezium and BN-KL, I plunged into the molecular cloud. My immediate gut feeling was to wish I hadn't. There is little so nerve-wracking as flying through a molecular cloud at high speed. As often as my technology performs flawlessly, there is always the slight doubt in the pit of my stomach that *Rocinante's* protective shielding will hold. Early airline passengers must have felt the same way. It's not the speed that bothers me; it's all that stuff coming at me at some tiny fraction less than the speed of light. I'm always aware how much of a punch it packs. I suppose the feeling is compounded by the fact that it is hardly possible to see anything because of all the dust.

To spare you from sharing my anxiety, let me tell you a little bit about how *Rocinante's* shielding works. It should come as no surprise that when I zip through interstellar matter at high speed, it doesn't see me coming. Of course, I don't mean "see" in the sense of perceiving my running lights. As fast as I can travel, light always travels faster and can run ahead. What I mean is that there is no mechanical warning of my proximity, no shove that tells the undisturbed gas to get out of the way before I arrive. My craft packs an incredible sonic boom, sweeping up everything in its path with a tremendous shock wave. A sheath of superheated, superpressurized gas is thrown up against *Rocinante's* skin, and it is against this that I need protection.

The pressure is not the main problem. It does increase stiffly as I approach the speed of light as measured by my Shangri-La factor, the ratio by which time passes more slowly in my craft than on Earth. For each doubling of this factor, the pressure quadruples. But interstellar matter is so sparse that even with these huge amplifications, the forces are easily parried, provided certain precautions are taken. When I traveled to the Milky Way's center, for example, I avoided all regions filled with more than one hydrogen atom in every cubic centimeter. At my peak Shangri-La factor of 13,000, I faced maximum pressures of barely 1 Earth atmosphere—easily withstood. My trip to Orion was rather leisurely, by comparison, because I had only to move 5000 light-years or so, and my Shangri-La factor never exceeded a few thousand. By the time I entered the molecular cloud, my craft was well into its deceleration phase and the pressures encountered were truly negligible, even given that hydrogen concentrations as high as 1000 per cubic centimeter, or more, were unavoidable in this comparatively dense environment.

I am more worried about *Rocinante's* skin getting too hot. To particles hitting it at speeds within a hair of the speed of light, my vessel's skin is as porous as a sponge. Oncoming electrons and ions could penetrate to depths of many centimeters (meters, even!), which would not pose a problem (*Rocinante* has a thick skin) if only they didn't also deposit all their enormous energies subcutaneously. My main defense is a powerful magnetic barrier that deflects the oncoming particles before they hit. Unfortunately, no magnetic shield is perfect, and on numerous occasions I have watched anxiously as blobs of plasma pierced the force field and struck home. The shield is also helpless to keep out particles of dust, and these have presented a steady, though lighter, onslaught. The surface of my craft, warming until it could radiate away the frictional heat, would reach temperatures of tens of thousands of degrees. No hard material—not even the ceramic of which *Rocinante's* shell is constructed—can survive at such temperatures, and I have watched nervously as patches of *Rocinante's* skin vaporized. I have had nightmares of my entire craft being eaten away, turning into a metallic/silicate steam and being sloughed off into space. But as you see, I am still here. What has saved me is that the evaporated ceramic forms an insulating layer. *Rocinante's* shape, and the play of pressures across it, holds the hot vapor in place; the layer of vaporized spacecraft skin, in turn, bears the brunt of the frictional heating and returns most of it to space with a searing radiance.

Rocinante continued to decelerate as we neared the Trapezium. I knew I was getting close to the cluster's illuminated cavity because conditions outside my craft had changed. Inside the molecular cloud I had encountered a ubiquitous infrared glow—detectable only with the correct viewing apparatus—that was created by warmed dust. Like any radiation emitted by a solid material, the color of this radiation revealed the dust's temperature. Now I noticed that, after an interminable stretch of dull sameness, the temperatures were starting to increase, the glow tilting toward shorter wavelengths. I was nearing a source of heat. Off to one side I saw a much hotter area, a few hundred degrees above absolute zero—the temperature of ordinary objects on Earth. I was passing by the BN-KL cluster, still hidden by dust. Straight ahead the gas seemed to be thinning slightly and warming still more. The Trapezium lay there.

Some turbulence, and a few roller-coaster swells, told me that I was crossing into a new zone. This transition was not the sharp shock I had been expecting. I knew that the massive young stars of the Trapezium emitted winds that sped outward at 2000 kilometers per second and carried nearly as much power as the stars emitted in light. When these winds hit the wall of dense gas that lay between me and the star cluster, they pushed on it with uncompensated force, compacting the exposed layers into the cold substratum through which I was now passing. At the same time, the intense ultraviolet rays from the stars fried the cloud wall, destroying the molecules and increasing the temperature 100-fold. The evaporation of this heated layer would have increased the pressure at the cloud surface still further, helping the winds plow their way into the dusty cloud. These were classic conditions for a shock wave: a piston of gas pushed into unsuspecting, cold matter, setting it suddenly into motion with an accompanying increase of temperature, pressure, and density. Such transitions were usually so sharp that as I had neared the expected location I had said to myself, "Don't blink"—I didn't want to miss it. I also gritted my teeth for a single, sharp jolt. But the transition turned out to be gradual, so gradual that I had time to analyze it and find out why.

The seemingly monolithic, gray medium I traversed had already revealed itself to be quite complex. All kinds of particles were present, and all were in motion. The molecules, of course, dominated. They, and the occasional single atoms, danced from collision to collision in straight lines, changing direction at random only as they bumped against one another. Their encounters often seemed amusingly like a square dancer's do-si-do, as the molecules looped around one another and atoms sometimes exchanged electrons gratuitously as they passed. Grains of dust—a trillion times heavier than a molecule—executed gently curving paths, seemingly oblivious to the frenetic small-scale action of the molecules. As they passed, the grains tweaked my electromagnetic sensors ever so slightly, indicating that they carried a slight electric charge. It was the charged grains' motion in the weak magnetic field, which I also detected, that curved their trajectories.

I puzzled over why the grains should be charged at all. Too few ultraviolet rays penetrated this far into the cloud to tear even a handful of electrons off the grain surface. In any case, that would have given the grains a positive charge, whereas they appeared to be negative. They must have acquired a few electrons, not lost them. Could they have acquired their charge by friction, like the static electricity on a rubber rod stroked by fur? Just then I noticed another component in the mix. In addition to molecules and whole atoms, there was a tiny admixture of ionized atoms and freely flying electrons. Curiously, the ions were not primarily those of the ubiquitous element hydrogen but were mainly derived from the much rarer carbon. Focusing on the electrons, I recalled hearing how they could charge up a big obstacle like a dust grain. Because they were lighter than ions, they moved much more quickly and thus hit the grains more frequently. If only a few of them stuck, that's all it would take to give the grain a negative charge. I smiled at this tortuous chain of reasoning all adding up to the gentle swing of the grains' trajectories. My instinctive first thoughts of fur, rubber rods, and static electricity no longer seemed so far-fetched. Using friction to remove electrons, after all, was no stranger than using random collisions to acquire them.

The very presence of freely flying electrons and ions, however, posed another mystery. How did they get here? I was still deep in the molecular cloud, well shielded from all nearby sources of ultraviolet radiation. The collisions between atoms were too gentle to knock them apart, but some agent had to be doing it. I gradually began to perceive yet another ingredient in this rich stew of particles. A tiny, tiny fraction of the ions and electrons were whizzing through the cloud with enormous random speeds almost indistinguishable from the speed of light. They were moving so rapidly (like the particles swept up by my craft at high Shangri-La factor) that they could penetrate the entire cloud. My colleagues called these cosmic rays, and I had encountered them before, in open stretches of the Galaxy. I was at first surprised to see them here, but what was to stop them from penetrating into every nook and cranny? These were the culprits that could collide with carbon atoms so forcefully that they knocked off an electron or two. But why carbon, rather than the much more common hydrogen? That was simple. Carbon held on to its outermost electron more loosely than did hydrogen. Easier to ionize, I recalled.

I could now see why this transition, from quiet cloud interior to raucous surface layer, was so gradual. The crushing impulse from the surface of the cloud was not being carried equally by all the particles. Near the cloud's surface, where everything was ionized, the motions of nearly all components of the gas were heavily regulated by the magnetic field. In the presence of a magnetic field, charged particles—such as ions and electrons—are thrown off their straight-line paths. The magnetism forces them into gyrations, and it is all they can do to spiral up and down the magnetic lines of force, wrapping coils around them like a Slinky. This means that it is the magnetic field that receives any impulse of momentum carried by the ionized particles, and it is the magnetic field that transports this impulse deep into the molecular cloud.

But there's the rub, literally. Deep inside the cloud, few of the particles are charged. Molecules and atoms abound, but they are not ionized and therefore are not affected by the magnetic field. Therefore, the magnetic field has trouble transmitting its impulse to the cloud's interior and passing it on to the particles there. The few electrons and ions, gamely tied to the magnetic lines of force, are the keepers of the cloud-crushing impulse. Occasionally, an ion collides with an atom or molecule or merges with a suitable electron and joins the ranks of the whole atoms. Only then does it give up its part of the cloud-crushing force and signal to the cloud's interior that powerful events are taking place nearby. This is a painstaking, gradual process and hence a gradual transition, collision by collision, from cloud to cavity.

Finally, I had passed into the outer layers of the cloud. There was no question now that my environment was under direct influence of the still-obscured stars. As I scanned my radio and infrared sensors, I could tell that the composition of my surroundings was changing. The largest, most fragile molecules had all but disappeared, leaving mainly robust carbon monoxide and molecular hydrogen intact. Then these gradually vanished, knocked apart by a combination of more violent collisions (the result of steadily increasing temperature) and the gradual increase in penetrating radiation.

The infrared glow ahead of me brightened. Then visible light, at first with a reddish cast and then successively melting into yellow and blues, bathed my craft with ever-increasing intensity. My image of the four bright Trapezium stars grew blinding, as I moved closer through the veils of dusty haze. The cluster's lopsided quadrangular pattern spread across a larger and larger portion of my visual field. Now the gas around me was visibly fluorescing, with its mix of atomic spectral colors. One final layer, an intense field of the pink light of hydrogen, and then an unbearable ultraviolet glare swept over everything, and I emerged from the cloud. I was in the cavity of the Trapezium.

15

Trapezium

If one tried to draw obvious comparisons between the environment of the Trapezium and that of the cluster at the center of the Milky Way, the former would be found wanting. First, one would have to imagine away the big black hole—there is none in Orion. Only four hot, massive stars made up the bright core of this cluster (I later found out that the nearby BN-KL cluster was richer in this regard), a far cry from the thousands I found in the Galaxy's center. And the stars here were sauntering about at measly speeds no greater than a few kilometers per second, compared to the hundreds of kilometers per second (influenced, of course, by the black hole's gravity) at which they move in the Galaxy's nucleus.

But here in Orion, there were contrasts and stark juxtapositions of structure that, in certain respects, surpassed those in the center of the Milky Way. I emerged through the ionized cloud wall at its closest point to the cluster, less than a light-year from the brightest star in the Trapezium. From this distance, 10,000 times farther than the Earth is from the Sun, the lead star was a pinpoint only 20 times brighter than a full Moon, and the other three Trapezium stars were considerably fainter than that. Yet unlike the Moon (or the direct light from the Sun, for that matter), these stars emitted most of their light in the ultraviolet part of the spectrum. The penetrating glare (even with shielding in place) was hard to take, and I quickly skimmed along the cloud wall to get out from the narrow gap between the star and the molecular cloud.

I now took in the scene from a more comfortable vantage point. The quartet of bright stars seemed to float in front of an endless wall of glowing pink. The hydrogen atoms producing this light were being dismembered by the impacts of ultraviolet photons, only to recover their electrons quickly and then have the process repeat itself almost immediately. Minor shadows in the ultraviolet bath, created by clumps of dust and indentations in the wall, were amplified by the atoms' sensitive response to light, creating a three-dimensional mottled appearance of curtains and billowing waves. In many places within the cavity and along the wall, the gas was set into motion, with ripples and shock waves creating their own light show of disturbed ions and atoms, in an array of colors. All of this had been visible from Earth. What had not been apparent was that the Orion Nebula was, for the most part, just a thin veneer lying behind the Trapezium. The entire depth of the pink-glowing screen was merely a sixth of a light-year, and behind it lay the vast, dark molecular cloud I had just traversed. I tried to orient myself in order to pick out features of the nebula that were familiar from Earth. I could visualize its appearance in a telescope as resembling a folding fan, its boundaries feathery and indistinct along a third of a circle, where the accordioned paper was unfolded, but angular and almost straight along the two enclosing arms. With considerable difficulty I deduced that one of those arms was a sharp, dark boundary, the silhouette of the foreground molecular cloud. The other, a bright bar that one could pick out by eye with even a small telescope, was apparently an illusion, a fold in the glowing pink sheet that observers on Earth happened to see edge-on.

As I moved farther away from the irradiated wall, in the direction of Earth, I could see that the nebula was nestled in the crook between two dense clouds, both of them cold and heavy with molecules. On the side of the cluster toward the Earth there was little molecular gas, but I was still not in open interstellar space. The cavity surrounding the Trapezium cluster was tenuous, the product of a multiplicity of colliding stellar winds. Only wisps of luminous nebular gas survived in this region—in most of the volume the gas was too hot. But a thin, dense shell, consisting of a mixture of atomic and ionized gas, surrounded the cluster on the sides not bordered by the molecular walls and gradually expanded away from it.

I suddenly realized that the Trapezium cluster was more than just a convenient light source that happened to illuminate the molecular cloud and make it a nice sight for amateur astronomers on Earth. It played an integral part in the fate of the molecular cloud. It owed its existence to the cloud. And, ungratefully, it was doing its best to destroy the cloud. The lid of atomic gas on the Earthward side of the Orion Nebula was being pushed away into space by the radiant heating and fast winds of the Trapezium stars. The same processes were evaporating the sheet of ionized gas overlying the molecular cloud. I ran the movie backwards to visualize what the scene must have looked like in the past—say, a million years ago—and realized that the Trapezium was not nestled into its cloudy nook by chance. Rather, it had created its nest by eroding the molecular gas around it. A few million years ago it would have been surrounded by the molecular cloud, completely embedded in it and invisible except for its infrared signature, much as the BN-KL cluster was now.

So entranced had I been by the nebular tracery that I had paid little attention to the star cluster and its role in shaping its environment. Now I had a closer, more critical look. The Trapezium cluster consisted of more than just its four bright stars. From my perch 50 light-years out, I could count more than 100 stars that seemed to belong to the cluster. The brightest ones were the most massive and would have the shortest lifetimes. Judging from their temperatures, these stars could not have formed more than 2 or 3 million years ago. But what about the less massive stars, which could look forward to much longer lifetimes? Did they, too, have to be so young? The stars' slight motions, which I had ignored until now, provided the crucial clue. Even though these motions seemed insignificant, just a few kilometers every second, they were still large enough to overcome the mutual gravitational attractions of the stars for one another. Unlike the cluster at the center of the Milky Way, whose high-speed stars would be tied together indefinitely by gravity, the Trapezium was ephemeral. A few million years hence and the stars will have drifted apart. Their current proximity to one another meant that all of the stars had to have formed en masse. They were all young.

A curious coincidence, I thought. As it was forming 2 or 3 million years ago, the cluster would have been embedded just beneath the surface of the molecular cloud. Had it been embedded more deeply, it would not have emerged yet. BN-KL was still buried, after all. Maybe it was a lucky break that the Trapezium was able to burrow out of its birthplace, to shine for Earth to see. But sheer coincidence was difficult to defend in this case. BN-KL was, in fact, very close to the surface of the cloud, and a census of its stars indicated that it was even younger than the Trapezium. Thus, in 2 or 3 million years, the BN-KL cluster probably will have eaten its way out of the womb, just like the Trapezium. No, the relationship of age to depth inside the cloud seemed a deliberate trend. But what effects could conspire, not only to make stars form in clusters, but also to make them form just inside the surfaces of molecular clouds? Why not in those clouds' deep interiors?

I scanned the skies looking outward from the cloud. If I consciously sought a clue (and I'm not sure I had any well-defined purpose), it was just a wild hunch that I would find it by looking *away* from the nebula. And I found something interesting: a circumstantial clue, to be sure, but one that reinforced my suspicions. Beyond the Trapezium, maybe 200 or 300 light-years away, was another young cluster, and beyond it a similar distance was a third. I remembered that I had seen the third one peeking around the opaque cloud's edge as I first approached the Orion molecular cloud from the other side. Neither of these clusters appeared to be as young as the Trapezium, if the absence of comparably bright stars was anything to go by. (One could never be certain that massive stars had ever formed in these clusters, but it seemed a good bet.) It also fit that each of these clusters was spread out over a much wider area than the Trapezium cluster, as though their stars had been drifting away from the point of common origin for a much longer time. The two methods of estimating age gave consistent results. I guessed the nearer cluster to be about 7 million years old and the more distant one to be celebrating its 12 or 13 millionth birthday.

It was not hard to imagine that this had something to do with the birth of stars and its relationship to the edges of molecular clouds. The progression was unmistakable. The farther a cluster was from the current wall of molecular gas, the older and more spread out it was. This suggested that the formation of the clusters was indeed related to the retreat of the molecular cloud and did occur near the cloud's edge. The molecular cloud was not just a receding glacier, exposing a moraine of stars as it retreated. The retreat of the molecular cloud was a sign that stars were being vigorously created out of its vast reservoirs of dust and gas. Even more remarkably, once the process got going near one wall of the cloud, it kept renewing itself with successive waves of star formation. All that remained was to figure out how.

I needed to find a trigger. The gas in a molecular cloud was evidently sitting precariously on the brink, almost ready to crystallize into thousands of stars but not quite able to take the plunge. It needed a shove to nudge it over the edge—a falling domino of star formation at Point A that would cause the molecular gas at Points B, C, and beyond to collapse in sequence. I had assumed that the process of star formation would be solely the province of gravity. But gravity alone proved to operate in too leisurely a fashion to create simultaneously the large numbers of new stars (not to mention the few massive ones) represented by these clusters. It must take an extra squeeze to tip a cloud into wholesale collapse under its own self-attraction. While I pondered the source of this compression, *Rocinante* drifted across a shock front that was much more vigorous than any of those I had encountered inside the Orion molecular cloud. My craft rocked sideways as it was pummeled by the wind of one of the Trapezium's hot stars, and I saw the pressure outside jump sharply. I had my answer: The massive stars provided the squeeze themselves.

My earlier impulse had been to attribute to the stars the warm and fuzzy attributes of a family, as though they were members of a species eager to reproduce. But now I saw they were really more like conquistadors, triggering and organizing star formation not in their own dominions but rather in the neighboring provinces, tens of light-years away. Star formation marched along inexorably, driven by the violent tendencies of the hottest few stars. The most massive stars in each cluster were the alpha males, shedding highly pressurized winds constantly through their brief but brilliant lives and then topping the winds off with violent explosions as they became supernovae. After 5 million years or so, the combined blasts of these select stars would sweep over a previously quiet piece of the molecular cloud, subjecting it to enormous pressures. Mildly dense patches of gas, which had been in no hurry to attain starhood, would implode, while new concentrations of matter would collect where none had existed before. It was not the only route by which stars formed, as I was soon to learn, but it seemed the most plausible way to create the sequential waves of rapid star formation that had eaten away so much of the molecular cloud.

The paradoxical aspect of this picture appealed to me. Whereas the hot stellar blasts seeded neighboring territories, the expedition was deadly on the home front. Local concentrations of gas, poised to create stars, must have been shredded, literally blasted out of their nests. The supernovae and hot winds cleared out the leftover gas in the home cluster, quenching star formation locally, while triggering it in some previously quiet locale 20 or 30 light-years away.

No supernova went off while I stared at the Trapezium, and even if one had, its connection to the next great wave of star formation would have been difficult if not impossible to prove. I decided to consign the whole concept of sequential star formation to the realm of promising hypotheses. There was a gap in my experience that was troubling me more. I hadn't seen a single star actually in the process of formation, much less a whole cluster of stars. Only gradually did it dawn on me that stars—little ones, for the most part—were forming all around me. And I didn't have to squint at the Trapezium to find them.

16

Points of Darkness, Shafts of Light

They had been there all the time. I had passed dozens of them at fairly close range as I sped through the molecular cloud, but they hadn't registered, I suppose, because I hadn't been looking for them. In retrospect, it's hard to believe that I saw the molecular gas as a monolithic sludge, rather than as the choppy and deeply textured medium it really was. Funny how perception can be conditioned by expectations. I had focused single-mindedly on reaching the massive stars of the Trapezium and must have been blinded by their brilliance, because I scarcely noticed the low-mass stars at first, even when I floated with them in the open cavity cleared by the hot stars' winds. Yet the little stars outnumbered the four brightest stars more than 100 to 1, and clearly had formed as part of the cluster. They were also forming, alone or in meager groups, everywhere else, from the depths of the molecular cloud to the hostile oven of the Trapezium's extended cavity. Finally I saw the knots of gas, dotted around. They were most obvious as opaque black globules against the luminous background of the nebular veneer. How could I have been so oblivious to them? Maybe they had simply been too difficult to detect in the molecular pea soup, far beneath the glowing wall. But as I re-entered the molecular cloud to check on what I'd missed, they stood out as islands—some light, some dark—in the infrared glow.

The more I looked, the more structured the gas appeared to be. This was not what I had expected. When I crossed the Milky Way I had seen "weather," to be sure—billowy interstellar cumulus, striated Galactic cirrus, even nimbus cloud decks. These clouds represented changes in temperature, even sharp boundaries, but they had all existed within prescribed bounds of contrast. The interstellar structures were expansive, their sizes typically measured in light-years. But these dark, dusty molecular knots were so cold, so tightly compressed that they seemed to be closing in on themselves, perhaps harboring some secret.

If the secret had been no more than the story of how such condensations could have formed, it would have been worth extracting. I was well aware of the obstacles that had to be surmounted in order to form a tightly compressed mass even from the dense, cold substance of a molecular cloud. True, gravity's imperative was to draw matter together, but I had already seen how rotation—angular momentum—could get in the way. Unlike the matter supplied by an orbiting stellar companion to a black hole or neutron star (a Cygnus X-1 or SS 433, for example), the matter that might ultimately condense into a solitary star need not rotate intensively from the start. But rotate it would, sooner or later, as gravity drew it together. Manifest in even the slightest variations of density, speed of drifting matter, and pressure in the molecular cloud were asymmetries that would inevitably amplify into a whirlpool of motion that would be sufficient to prevent wholesale collapse.

My concerns seemed to be borne out when I got close enough to several of these clumps to observe their structures. They were flattened (usually a reliable indicator of rotation), and my Doppler measurements detected the signature of rotation, as well: blueshift due to gas coming toward me on one side, redshift due to receding matter on the other side. But I soon began to doubt that the action of rotation was the only effect responsible for the wafer-like structure of these clumps. I didn't have to make any special efforts to discover that the magnetic field was organized in a way that was quite different from that in any of the rotating disks I had studied before. I could trace the magnetic field because of its strange effects on matter, turning it into springy toffee, extruding it into filaments and streamers. The streamers of gas, in turn, outlined the magnetic lines of force, just as the iron filings that I used to play with as a child could outline the lines of force created by a bar magnet. I had encountered the cosmic visualization of magnetic field lines even before I had left Earth, in the prominences and streamers that frequently erupted from the Sun. I had seen similar arcs of magnetic field erupting from the accretion disks in Cygnus X-1 and SS 433 and even emerging from the anemic swirls of gas near the black hole in the Milky Way's nucleus. But those were highly chaotic situations, in which the magnetic streamers never stood still. They were tousled and unruly, unpredictable and turbulent, and often explosive.

Here everything was much more calmly organized. Each wafer was wrapped with combed lines of force, some of which seemed to be matted along its surface. Others protruded like porcupine quills, except that instead of ending in sharp bristles, they spread out and joined smoothly onto magnetic lines of force in the surrounding, uncondensed gas. It was as though these collapsed clumps of gas, having fallen partway in on themselves, could not quite bear to take the final step of detaching themselves fully from the medium that had given them birth.

The elasticity of the magnetic field endowed these knots with a kind of liveliness. It was not only the rotation that was giving pause to their contraction; they were also tethered to the magnetic lines, which seemed to be snapping them back from the brink of collapsing too quickly. They jiggled and rocked, as though they were bouncing in a cradle of bungee cords, yet there was no doubt about the inevitability of their contraction. As the knots shrank, their rotational motions became more evident and eventually rivaled the magnetic tension in attempting to stymie gravity. Rotation, too, pulled on the magnetic field lines, twisting them. I remembered my old lessons about how magnetic lines of force would often appear to be embedded in gas and to move with it, like stripes in salt-water taffy. Thus it was no surprise to see the magnetic lines pinching inward as the knots sank under their own weight and curving into lazy spirals as the angular momentum began to spin them around. But as the gas in these knots became more condensed, it looked as though their hold on the magnetic field was slipping.

The effect was subtle, at first. I noticed that the pinching seemed not quite so strong, the curvature not so great, as my intuition had led me to expect. The rotation increased inexorably, but the twist in the curves of magnetic field did not keep pace. It was not that the magnetic field was being left behind without a struggle. There was a price to pay, a drag on the rotation that left the spin just slightly less than it would have been had angular momentum been able to do its work unfettered. It was clear why this was happening. As the gas contracted and became denser, its particles closer enough together, it was also becoming colder. The few atoms that had lost electrons, for whatever reason, greedily retrieved them. This doomed magnetism's grasp on a clump, because only the particles with electric charge, the ions and electrons—not the ordinary atoms or molecules—"felt" the magnetic field. Those few particles alone had shouldered the entire burden of carrying the magnetic field, making it twist and compress as the gas rotated and shrank. Now, with these particles becoming rarer and rarer, the burden of reining in the magnetic field was just too overwhelming, and the field slipped away.

At the same time as the magnetic yolk was slipping free, it was taking its most serious toll on the clump's rotation. So much of the rotation had been shed that the (now much rounder) globe could begin serious contraction. It pulled more tightly together, its surface glowing more brightly as the internal pressure grew. It began to heat up again. I estimated that the temperature of one clump's core was approaching 100,000 degrees, and it was only a matter of time before it would reach 1 million degrees, the threshold for thermonuclear reactions to start. This would not be the powerful transformation of four hydrogen nuclei into one helium, the reaction that powers the Sun and most stars, but a milder one that initiated formerly benign (cold, dusty, passive. . .) pieces of molecular cloud into the shining mysteries of active starhood. There was now no doubt that this clump of gas was destined to become a star.

Without the magnetic coupling to resist its rotation, the clump's outer layers spun faster as it continued to contract, and as more matter fell into its grasp it once again developed a flattened shape. This time the flattening was much more pronounced, a bulbous core surrounded by a thin disk like the planet Saturn. Then the nuclear initiation began. Deep in the core, nuclei of that rare isotope of hydrogen, deuterium (a duo consisting of one proton and one neutron) grew hot enough and came together with enough force to obviate Nature's safeguards against nuclear power—the protons' electrical repulsion—and fused. A wave of turbulent energy spread outward through what I may now be justified in calling the "protostar." The protostar's outer shield of molecular gas, matter still drifting in, unaware of the developments below, was presumably overtaken by this reactive wave and shocked into retreat.

I say "presumably," because I reconstructed these final events in retrospect. Before the wave of heat had swept over my craft from the newly assertive protostar, I was distracted by something that was of little importance in itself, but showed me the way to the next stage of star formation. I was suddenly aware of a small, tattered bit of glowing nebula coming into my field of view. It was very bright, and from the pattern of spectral colors I could tell that it was not just a disembodied bit of the main Orion nebula, fluorescing under the ultraviolet glare of its hot young stars. This gas was glowing because its atoms were being slammed against one another: It was a fragmentary shock wave, seemingly come out of nowhere. I stopped my slow drift and was now virtually stationary with respect to Orion's molecular clouds, but this stuff was coming at me at surprisingly high speed (even if I accepted that this was a shock wave), closing in at several hundred kilometers per second. I instinctively looked out my opposite porthole to check its destination and was astonished to see another bright patch, this one a quarter of a light-year off (and therefore appearing much smaller, but otherwise similar) and moving away from me at a similar speed. Beyond that, I saw a string of more distant splotches, a whole train of these disembodied shock waves.

They were all lined up; it was obvious that they were related. But where were they coming from? It didn't take me long to find out. At this moment, I crossed their path and felt a sudden impulse pushing *Rocinante* in the direction in which the knots were traveling. So they were not completely disembodied, after all. There seemed to be a continuous, fast stream of gas that connected them. It wasn't a steady stream. Its speed and intensity fluctuated, which meant that the faster portions could catch up to the slower segments and ram into them. This game of tag was what led to the shock waves and the violent atomic collisions that engendered the intense colored light.

I followed the stream back toward its source. Subconsciously, I must have expected to revisit one of the dramatic environments of my earlier travels—a jet from a black hole or neutron star binary, perhaps—despite the fact that this stream of gas was traveling a hundred times slower, merely a few hundred kilometers per second. It was a bigger surprise than it should have been, therefore, to find that I was being led back to an object very similar to the one I had just left. As I approached, I saw the familiar thin, dusty disk surrounding a glowing sphere. This protostar seemed to have evolved to a slightly more advanced stage than its compatriot. The contracting envelope had been, more or less, entirely dispersed, the sphere was slightly smaller and hotter, and the disk was just a bit more spread out.

With difficulty, I could trace the jet nearly all the way to the protostar. It was faint here, the bright splotches not having formed this close to the jet's source. The jet seemed to emerge from the very innermost regions of the disk, just outside the point where it brushed the globe of the newly radiant star. Although strong outward propulsion dominated its motion, the helical shape of the magnetic streamers that embraced it revealed that the jet also swirled about its axis. This was not surprising, given the jet's origin in the rapidly rotating disk. What did surprise me was to see, once again, the streamers of magnetic field, this time outlining the jet like a candy cane. The swirling jet seemed to be wrapping lines of magnetic force tightly about itself. Or was it the other way around? I suspected that it was the jet that was doing the bidding of the magnetic field. The magnetic field here was strong, much stronger than the field of the protostar I had visited previously. Probably it had been amplified by the rotation of the disk, which had become hot enough to hold onto its magnetic field once again. As I had found in my earlier travels, energy was fungible: It could slip just as easily from its rotational form to its magnetic form as from rotation to heat or from heat to light. The field was wound up like a mainspring, and I could sense its tension as it struggled to unleash its pent-up energy—to convert itself back to the energy of motion. But magnetism's peculiar form of tension would not permit it to lash out equally in all directions. As it propelled the jet forward, the helical turns of magnetic field squeezed inward toward the common axis, seeming to suffocate the jet like a boa constrictor. This seemed likely to be the cause of the jet's slenderness, yet the constrictive and propulsive forces were not finely balanced. The jet lurched forward in spurts, its direction wavering and its degree of focus equally unsteady. As I peered outward along the jet, in the dim distance I could make out these spurts colliding and shocking, to create the first of the bright splotches that had given the jet away.

As I pulled farther from the center of the disk, the scene became much more familiar. If the magnetic field had seemed combed out and orderly during the early stages of the previous protostar's contraction, by this stage in the evolution of the present protostar it had assumed a disorderly, chaotic structure.

Flares erupted from this disk, reminiscent of the ones I had seen in the accretion disks of Cygnus X-1 and SS 433. Although the most powerful and fastest expulsions of matter came from the center of the disk, where the effects of both gravity and rotation were most pronounced, outbursts were occurring all along the surface of the disk. How strange to see the fireworks that I had come to associate with black holes and neutron stars repeated, in milder form, as gravity and rotation struggled over a mundane prize like a young star.

As had been almost universally the case whenever I encountered a new phenomenon, this protostar's jet was no freak. As I looked about me, I could now recognize dozens of trains of these shock-splotches strewn about the sky, a network of shafts like so many pick-up sticks. It was no surprise that jets came in symmetrical pairs, emerging in opposite directions along the axis of rotation of each disk. It seemed that virtually every newly formed star, heavy or light, produced a pair of jets. If I hadn't noticed a jet emanating from the first protostar I had visited, it was only for want of bright tracers. Had I accidentally wandered into the beam, I surely would have been apprised, rudely, of the jet's existence. Now, peering toward that first protostar from a distance, I could see its train of aligned shock waves clearly.

I gradually began to pull away from the Orion region. As my field of view expanded, so did my understanding. I realized that Orion was much more than a backdrop for its main attractions, the Trapezium and the Nebula. In every quarter it was a beehive of activity: stars forming everywhere, their jets, winds, and radiant output churning up the molecular gas, modifying their environments, creating the conditions for further creation. Orion was a venue different from the others I had visited. The Crabs, the cannibalistic black holes Cygnus X-1 and (perhaps) SS 433, even the Galactic Center—they were all definite places, coordinates one could dial into one's navigation system. When I set my sight on Orion, I had had in mind an excursion to—where? The Nebula? That insubstantial glowing sheet? The Trapezium? Just those four stars that happen to be this week's centerpiece? Next week (of course I mean next millennium, or sometime in the next million years) some other stars will have stolen the spotlight, illuminating this ponderous mass of dust and molecules from a different angle. And someday this entire complex of molecular clouds will be gone, and a new Orion will shine somewhere else in the Galaxy.

At this seemingly unexceptional moment, my perspective on the voyage suddenly changed. One cannot regard Orion as a "place" to visit, any more than one can regard "North America" as the destination for a two-week vacation. Suddenly, I saw my trip not as a sequence of visits to curiosities, each with its set of mechanisms—those of gravity, motion, and equilibrium—but as an exploration of *process*. Gravity drew matter in toward the center of a galaxy. The Crab pulsar was locked in rigor mortis in the aftermath of a star's violent death. And Orion's symbolism was the sweetest of all I had seen so far: The birth of structure from chaos.

I cannot explain why, but my attention was suddenly diverted to an unimposing swarm of condensations. They were like so many others I had seen, an unassuming cluster of disks, but there was something different about them. I interrupted my departure, wheeled about, and plunged back into the molecular cloud. It turned out that I was responding to the call of the wild.

17

The Dust Storm

During one of my journeys before leaving Earth I encountered a sandstorm. I am sure it was not one of the more extreme versions of this phenomenon, but it was frightening enough, all the same. A sandstorm terrifies by choking all of one's senses. Smell and taste become indistinguishable and then disappear altogether. Sight, of course, is useless—better to protect one's eyes as best one can. The whir of sandblasting against car, fencepost, and clothing provides a percussion that accompanies the droning of the wind, masking all sounds that carry useful information. And touch signals only pulverized grit getting into everything.

Although I was not exposed to the (far harsher) elements outside my craft as I traveled into the disk, the environment in which I now found myself had similar psychological effects. If anything, it was even more disorienting. There was no ground beneath my craft, no porch railing to hang on to. A uniform blankness extended in all directions. I had lost my bearings the moment I descended into the dense layers of swirling dust, losing sight of the Trapezium's now familiar stars. From afar, the disk had seemed so intriguing, a dark teardrop silhouetted against the pink of the nebula, near the wall of the Trapezium's cavity. The blunt end of the teardrop faced the Trapezium's brightest star and was clearly being shaped by its windy punch. On the leeward side, the stream of dust being scoured off tapered down nearly to a point. The wind's impact was doing its best to decimate this dusty wafer, but I could easily see that its effects were destined to be superficial. The disk had already shrunk under its own gravity to such a degree that it was nearly impervious even to the buffeting of a hot star's wind.

It had beckoned as a respite from the stark glare of the brightest of the Trapezium stars, and that, in a sense, it was. As soon as I entered its outer fringes, the harsh light mellowed. First I lost the ultraviolet glare—a most welcome relief. Then the blues went, leaving a yellowish cast. This steadily reddened as I sank deeper toward the tight plane along which most of the dust was concentrated. I knew that this disk was orbiting a newly formed star; it was in fact the debris left over from the star's formation. This was not a massive star, like the Trapezium behemoths, but one that would probably come to resemble the Sun when it reached maturity. The star poked above the disk—indeed, it had already cleared a hole around itself—and I had a chance to study it as I approached. From the star's color, its size, and the intensity of its light, I could surmise that it was still heating up, which meant that it was not through shrinking down toward its final size. But neither was it falling in on itself. To support its weight and slow its shrinkage to an imperceptible rate, the temperature in the center of its core must already have risen above 10 million degrees. This meant that its nuclear-powered furnace had sputtered into action and was ramping up to the level at which it could sustain the full-fledged star for billions of years. It was already a self-sufficient body; the disk seemed superfluous.

Yet it was the disk, not the fledgling star, that had drawn me here. Such dusty disks of debris contained the only possible raw material from which planets could have formed. I needed to see the process for myself, and preferably around a star with which I could identify—I mean, of course, a star like the Sun. Witnessing stars themselves being born had extended my appreciation of phenomena that I had encountered elsewhere. The relationships were forming themselves into a mantra in my thoughts. Gravity's inevitable tendency was to draw matter toward a common center. Heat and spin hampered such condensation, but when these were disposed of—often (I had learned) through spectacular jets of gas or the springy dynamics of magnetic fields, or both—gravity would nearly always win. Under much more extreme conditions, near black holes, I had seen how gravity could endow matter with violent motions and unleash torrents of energy. If that had seemed to symbolize a destructive side of gravitational force, then lately, in the stellar nurseries of Orion, I had discovered its creative side. Gravity liked texture. It was inclined to produce interesting patterns (the spiral arms of the Milky Way came to mind) and to create glowing new structures from featureless lumps of gas.

But planet formation seemed different. It had to occur after the star had formed, or at least late in the star's process of condensation. Planets were, after all, mere afterthoughts to stars, leftovers that had congealed into bodies that sometimes—at least once, but who knows how commonly—hosted life. Were they really so different from the little cinnamon crusts that my mother had baked from leftover strips of dough when the main event—the pie—was cooling? By that reckoning, planets could seem so inconsequential. Yet I remembered those crusts as the real treat on baking days. Little as they represented in terms of matter and energy, planets would always be regarded as special, even by the most hard-nosed astrophysicist. They were irresistible objects of intrigue mainly because they reminded us of home—they were all potential habitats. But from a different perspective, they were special in their own right. One could think of planets as distillates of the rarest elements, of a purity not found elsewhere in the cosmos. Hydrogen and helium having been boiled off, and metallic solids separated from brittle ceramic and glassy minerals, the construction of a planet simulated the archetypal labors of smelting, metalworking, and pottery. Amid the unyielding and often crude chains of events that shaped the Universe, it seemed surprising that structures so delicate should emerge spontaneously. Even planets like Jupiter, whose hydrogen-rich sheath and odd luminous emanations made many astronomers consider it partway toward being a star, had its rocky core.

The disk I was exploring seemed very far from evoking the comforting terra firma of a planetary surface, however. If ever one sought a visceral representation of chaos, filled with neither darkness nor light, lacking perceptible structure and orientation, the deep interior of such a disk would be a good candidate. I wondered whether the author of Genesis had preceded me to this spot. How does one go from this featureless morass to the rust-red globe of a Mars, a candy-striped Jupiter with its Great Red Hurricane, or a blue- and white-speckled Earth? I thought of the punch line to that ancient joke about asking directions in the countryside: "You can't get there from here." But clearly you could, and I needed to find out how.

Objectively, the scene was not particularly violent, at least by the standards of my previous travels. If anything, the calm quality of this disk added to my unease. I had earlier lost sight of the star at the center of the disk, its location finally marked by a fading infrared glow that was my last positional point of reference. I guided *Rocinante* into the dust plane and brought my craft as nearly as possible into synchronization with the orbital motion of the dust particles. This meant that I was orbiting the central star at a few kilometers per second, but my orbital motion was imperceptible because the dust that surrounded me was orbiting at exactly the same speed. The individual grains were not stationary with respect to my position, however. From my moving platform, I perceived them to be darting in completely random directions at about the speed of a fast-moving car. Though this was little more than a thousandth of the orbital speed, and was completely negligible compared to the speeds to which I had become accustomed in my travels, it was shocking to see particles of macroscopic size, that one could scoop up and hold in one's (well-gloved) hand, barreling along in this way. It was as though I were back in the sandstorm, with the grit particles moving around at high speeds but with no wind to drive them in a fixed direction.

What struck me as most strange was that I was suddenly immersed in an environment that was comprehensible to human senses. Everywhere else I had encountered bodies so large (or atoms so small) and speeds so incomprehensible that I had been forced to abstract an idea of their dimensions through the use of my instruments or the conscious interpretation of my observations. I knew that the atoms of air at room temperature were moving 10 times faster than these grains, but because I couldn't see the atoms I had never had to address the reality of that motion. Here I could sense directly what was happening, to the extent of having to put up with the racket of these tiny grains striking *Rocinante's* windows dozens of times per second. They made the same kind of rattling noise that had driven me to distraction during that sandstorm many years ago. And through the windows I could see these particles flying at me from all directions, like flakes in an infernal snowstorm. I found it curious that a terrestrial sandstorm would linger in my memory as having temporarily robbed me of my senses, whereas its cosmic counterpart would suddenly restore them to relevance. I took this as a preliminary, indirect sign that perhaps some order amenable to life could emerge from this formless debris.

Not that I was surprised to see the grains in random motion. I reckoned there was enough unconsolidated dust to construct, maybe, a thousand Earths, spread in a disk that extended out from the central star to at least 100 times the size of the Earth's orbit around the Sun. The dust layer was thin, but not infinitely so. Most of the dust was concentrated along a sheet that was about 1000 times thinner than it was broad. Yet the breadth of the sheet was so great that it was still thick enough to hold a thousand Earths laid side by side. The disk's finite thickness went hand in hand with the random motions that had announced themselves so annoyingly along *Rocinante's* ceramic skin. The motions of the grains were exactly analogous to the motions of molecules that give pressure to a gas. If this pressure did not exist, the gravitational attraction of the central star would draw a gaseous disk down to a plane of paper thinness, and the same thing would happen to a disk of dust if the particles did not have some random motions superimposed on their orbital circulation. The average speed of the grains was linked to the thickness of the disk via the star's gravitational attraction, just as the speeds of gaseous atoms had been in proportion to the thickness of the disk I had visited in Cygnus X-1. I was comforted to find that this dust disk was a good facsimile of the gaseous disks I had seen time and again, only this time executed in the medium of pulverized stone.

What caused the grains to dance continually? Unlike the disks of gas I had visited before, there were no eruptions of magnetic flares, no bursts of radiation remotely intense enough to stir up these heavy particles of grit. Still, I surmised that this thickening was not accidental. Had the disk been thinner, the dust particles pressed more closely together, the gravitational attractions of the dust grains for one another would have partially overwhelmed their organized attraction to the central star. Once this happened, the smooth structure of the disk would dissolve into striations, braids, and swarms of dust that would not stand still. All of this activity would amount to a stirring action, which would give the dust even more random motions than I had measured. Thus it was gravity that must have stirred up the grains. Just enough stirring, and the dust's self-attraction would balance the pull of the star. It seemed like a new kind of equilibrium, one in which two sources of gravitational attraction—dust for dust versus star for dust—had achieved a delicate balance. But I quickly realized that it wasn't completely unfamiliar. A similar competition, in which stars played the role of dust grains and the Galaxy as a whole served as the central star, gave rise to the spiral waves that decorated the Milky Way. I was pleased to see the connection, but a little wistful that true novelties were becoming rarer as my travels continued.

At this point I was beginning to weary of the lack of texture in the background. Where were the planetary structures that were supposed to emerge from this soup? I didn't see anything of even the size of a marble. I peered more intently into the abyss, trying to spot some clue. Amid the dust I could see a distance of perhaps 10 Earth diameters, a mere hundred-thousandth of the disk's breadth and barely 1 percent of the distance to the disk's surface. By astronomical standards, this was the equivalent of pea-soup fog. Most of the dust particles in my near field of vision seemed to be only a little bit coarser than the dust I had seen elsewhere. It was still like fine soot, the typical grains having sizes only slightly larger than a micron, a ten-thousandth of an inch.

It suddenly struck me what was so puzzling about this aggregation of dust. Where was the gas that usually went along with it? I was well acquainted with dust as a routine component of the matter between the stars. Dust had accompanied me nearly everywhere I had visited. At the outset of my first journey, it had frustrated my attempts to see the center of the Milky Way. Only in the superheated regions of explosive bubbles had it been scarce. There I could understand that it had been evaporated by the impacts of the energized ions. Nearly everywhere else it was a trace component, barely 1 percent by mass. But here, it was the dominant material.

Those among you who have never left your home planet may be surprised to learn that one grows accustomed to thinking of all elements other than hydrogen and helium as luxuries. As I have already noted, from the inanimate perspective of the cosmos, planets are most distinguished by their immense concentrations of chemical elements such as oxygen, carbon, silicon, and iron. Nearly everywhere except in planets (and the interiors of some weird bodies known as white dwarfs), these elements taken together make up barely a percent of all matter by mass. It is no accident that this elemental fraction is similar to the fraction of matter that makes up the grains in the spaces between the stars. In those relatively cool environments, the heavy elements are able to exercise their natural tendencies to combine in certain ways and to condense into solids. On Earth, even those elements considered scarce—tin, platinum, and uranium, for example—are enormously concentrated compared to their abundances in space.

In this disk, it seemed that I was witnessing the early stages of this purification process. Somehow, the dust had been winnowed from the gas, and much of the latter blown away. Had the gas been pushed outward through the dusty matrix by gusts of wind from the youthful star? Or had the dust, striving to orbit against the friction of a warm, gaseous environment, drifted inward, leaving the gas behind? It was too late to reconstruct that episode of the story. Anyway, it didn't matter much where the gas had gone: It would quickly mix with and become indistinguishable from the ordinary matter of interstellar space. The important thing is that the planetary raw material had purified itself, concentrated just those elements that could provide the rigid framework of a habitable body if other conditions were right. I counted this as another sign that I was on a fruitful track.

I scooped up a bagful of grains and drew it inside *Rocinante* for closer examination. The interstellar dust I had encountered earlier, virtually everywhere except in the hottest environments, had been a fine, solid substance. There had been the odd clumps of particles stuck together, the very rare speck approaching the size of the point of a pin. This dust was noticeably different. It was crumbly. But that wasn't because it consisted of a different material. When I magnified it and examined its structure, I saw delicate filigrees that looked fragile and broke apart easily when flexed and yet exhibited a surprising toughness when I tried to crush it in my microscopic vise. These were composites of the ordinary grains I had seen before. An observation I had made casually, and to which I had paid little attention, suddenly assumed immense importance. These grains were bigger than the grains of interstellar space. They had started down the long path toward forming planetary structures, after all. I had arrived near the beginning of the process, when they had just begun to stick together. One would have thought that they would shatter, colliding at their random speeds of more than 100 kilometers per hour. Yet the successive impacts not only left them intact but seemed to cement them together. I laughed aloud in astonishment, but the main emotion I felt was awe at the thought that mighty planets like Jupiter could have grown from these tiny smudges of interstellar soot.

At the rate they were scooting across the disk, the smaller grains would run into a mate every year or so. Not every collision would lead to coalescence. Some glancing encounters would leave the grains little scathed. And a good fraction of the collisions, especially those involving the less consolidated clumps, would shatter the participants. But it was clear that things would develop quickly, and the grains would continue to grow, in spite of these inefficiencies. I started to pull *Rocinante* slightly out of the densest layer to get a longer view, at the same time edging closer to the central star where I thought events would be happening more rapidly. I settled in to watch, then . . . Bang! *Rocinante* was struck a blow out of all proportion to the annoying but harmless rattle I had endured up to that point. I checked for damage . . . negative. Then . . . Bang! again, only this time I was ready. I trapped the offending projectile: a small pebble barely a millimeter across. This was something you might not stop to remove from your shoe, but here it was a truly extraordinary creation, an object containing a billion of the granular building blocks. Stick a billion of these together and you'd get a meter-sized boulder. Another billion-fold and you would have a mountain. And an amalgam of a billion mountains . . . that would be enough to make an Earth.

Alerted by this harbinger of growth and change, I began to see things I had missed before. These much bigger grains were not so rare as I had thought, although they were much rarer than the tiny dust particles. Whereas the dust grains had peppered every inch or so of the space surrounding my craft, the pebbles were hundreds of meters apart. Their relative rarity explains why I had not noticed them before—that, and my good luck in having avoided a frontal collision before now. The pebbles already contained a few percent of the solid matter in the disk, and that percentage was destined to increase over the next thousand years or so, until they had swept up and incorporated most of the dust. Then, a traveler visiting the central layers of the disk might even be able to see out, to appreciate at once this proto-Solar System and its spectacular setting within the Trapezium's bubble.

Would I be that traveler? I had a choice. I could rev up *Rocinante's* thrusters to accelerate away, thus slowing down the passage of my time, with the intent of returning to witness the development of this system at a later stage. Or I could give up and return to the intense gaseous sprays of newly formed stars, which now looked surprisingly benign by comparison. One thing was clear: I could not remain embedded in this disk. It would be 10,000 years, or more, until enough of these small particles had coagulated so that I could navigate safely among the larger chunks, avoiding dangerous collisions. I pulled above the disk, back into the welcoming pink glow of Orion, to decide what to do.

18

The Shepherd

Serendipity—finding important things by chance—has always been one of the astronomer's best friends. By definition it is unpredictable, and its impact is often most poignant when conditions look least hopeful. It is safe to say that the most significant discoveries made by my generation of astrophysicists were serendipitous; the jets of SS 433 offer a good example. Astronomers looking for one thing found something entirely different, and in many cases they weren't even looking. The discovery figuratively descended from the sky and bopped them on the head. My brush with serendipity was not so dramatic. Faced with two unpalatable choices in my search for planets in formation, I unexpectedly found a third way.

Across a gap of what couldn't have been more than 2 or 3 light-years, and partially embedded in the wall of molecular gas behind the luminous façade of the Orion Nebula, I spied another disk that had the earmarks of dust but also exhibited some interesting features that were missing from my present venue. This disk was banded with dark and narrow concentric rings, where dust seemed to be absent. Like the rings of Saturn, I thought. During my childhood days as an amateur astronomer, it had been considered an easy test of visual acuity to spot the "division" in the rings that had first been noticed by Cassini in the seventeenth century. This could be done with a small telescope. If you had a bigger scope and a steady eye, you could find hints of the many other narrow gaps that observers had discovered over the years and that stood out prominently when the first close-up pictures came back from Pioneer and Voyager. It later turned out that Uranus, Neptune, and even Jupiter had rings, although these were far beyond the detection capabilities of amateurs with small telescopes. The latter systems were like negative images of Saturn's rings. Instead of dark gaps between bright annuli, the divisions consisted of narrow, bright rings of reflective particles separating broad, empty spaces.

A theory had been developed to explain both the gaps and the narrow rings. Neither had been expected, because even the small random motions of the particles (which would necessarily arise from the same kinds of gravitational effects that had generated them in the disk I had just visited) would cause narrow rings to spread and merge and gaps to fill in quickly. It was hypothesized that the rings were held in trim, and the gaps kept open and sharp, by small moons orbiting the planet. To clear out a gap, a single moon would have to orbit within it. And to channel particles into narrow rings, there would have to be a pair of moons, locked in synchronized orbits on either side. These moons didn't have to be very big; in fact, their existence had been predicted simply because their presence would explain the gaps and rings, long before they were seen. When they finally were discovered, it caused a sensation in the world of planetary studies. They were called "shepherd" moons, because they guided the streams of dust and debris and prevented them from getting out of line. "Sheepdog moons" might have been more appropriate, but the anthropomorphic name stuck.

I sped across the distance that separated me from this banded disk and hovered just above it. I knew instantly that it provided me with a new option, because it was obviously a planetary system caught at a later stage of development—just what I was looking for. The dark bands were indeed narrow gaps, and I counted on each of them being kept clear by a sizable shepherding body, in this case a small planet. The spaces between the gaps were still filled with debris, including some fine dust, although this time the dominant contents seemed to consist of an assortment of bodies ranging from millimeter-sized pebbles up to large rocks and mountainous fragments several kilometers across. These must have been built up by the same coagulation process I had witnessed earlier among much smaller particles. As these bodies grew, they swept up material at an ever-increasing pace, yet the time it took to accumulate a meter-sized boulder was impressive: tens of thousands of years. To build a smallish mountain required millions of years. The spaces filled with rubble looked as dangerous as ever, but now I had the option of riding in the clear area of a gap. However, before I risked my neck again by descending into the plane of the disk, I wanted to see the body that was alleged to be keeping the gap clear.

It took a while to find it. Avoiding the most concentrated layers of the disk, just in case, I selected a particularly broad and well-formed gap and circled above it. It wouldn't have worked for me to synchronize my orbit around the star with that of the nearby orbiting debris, because the purported "shepherd planet" would be orbiting at the same rate. Without exceptional luck I could get locked into a perpetual game of cat and mouse (or mouse and cat, given that my prey was much larger than I) and might never encounter the planet. I knew that tiny moons could shepherd large, prominent rings. But it is hard to appreciate just how powerful the shepherding effect can be until you see for yourself the discrepancy between the width of the gap and the size of the planet required to clear it. In fact, I passed by the planet twice before I even noticed it. I had expected at least to sense the tug of the planet's gravity as I passed by, but I must have been traveling too fast or have been too far above the disk for this to be noticeable. On my third time around, I spotted it visually. If it hadn't been for the corroboration of my calculations, I would have ignored it and kept on looking. From all appearances this was an insubstantial body, slightly more than a thousand kilometers across and weighing (by my estimates) only one-thousandth as much as the Earth. Yet the gravitational attraction of this planet, exerted steadily over thousands of years, had cleared the debris from a gap that was a thousand times wider than the body itself!

I closed in on the body and synchronized my orbit with its motion. It was not the kind of object I envisaged when I thought of a rocky planet. Planets like Earth, Mars, and Venus are highly structured. They are layered like an agate, their centers rich in iron and the heavier minerals, and their outer layers consisting of successively lighter materials until (in the case of Earth, at least) the top layer, the lithosphere, literally floats on the heavier mantle. From the gravitational tugs I finally measured as I approached, I could tell that this was still a mainly undifferentiated lump, reflecting its heritage as the sum of the countless agglomerations of dust, pebbles, and rocks that had gone into constructing it.

The collisions that had built up this body had not been entirely random, however. Objects as large as a few kilometers, or less, grow in a haphazard way, accreting whatever they happen to run into. But when any concretion within the disk reaches the size of a mountain, a dramatic change occurs in the way it interacts with its surroundings. The growth becomes more directed, and inexorable. Whereas before, the concretion merely ran into neighboring, smaller bodies by chance, its gravity now begins actively to focus nearby debris onto collision courses with it. Like some kind of inanimate Pied Piper, it develops a retinue of smaller bodies that slavishly trail behind it. As they jostle one another for position, some of them move too close and merge with the leader. This focusing effect greatly accelerates the rate at which a protoplanet can grow: To double a mountain-sized body would take millions of year, but To bring this shepherd planet from half its size to its present dimensions might have taken only a few hundred years, if that.

There is another intriguing aspect to this runaway process of accretion. Once a body has become the dominant protoplanet of its domain, the growth of competitors to this anointed one dies back. I ventured above the disk to look for other bodies of similar size: None were visible in the vicinity of this gap, though there would have been enough matter to create them. I saw plenty of objects measuring tens, even hundreds, of kilometers, but there was nothing in sight to challenge the dominance of my shepherd planet over the domain of its own gap.

Such a breakneck rate of growth cannot continue indefinitely. Eventually the protoplanet swallows everything within its reach and runs out of matter to accrete, even given its ever-increasing powers to attract smaller particles. This happens precisely when it has cleared out its gap, a threshold that depends on the mass of the planet and the thickness of the disk. This is why I found this planet to have grown tantalizingly close to an Earth-like mass (a factor of 1000 seemed awfully close, compared to the cosmic dust bunnies I had encountered earlier), but no closer.

Having convinced myself that it was safe, I maneuvered *Rocinante* into the gap and flew, in formation, ahead of the shepherd planet. It was thrilling to see the seething boulders, mountains, and pebbles held at bay along either wall of the gap. I thought of Moses parting the Red Sea and wondered whether the parallel walls of water had given a similar impression to the Israelites. I fantasized that somehow it was my gravity that had cleared the way, exploiting some amazingly subtle interplay of forces. If this sounds like hubris, it probably was, for at that moment I suffered the kind of shock that often befalls those who overestimate their ability to predict (much less to control) the forces of nature.

I noticed a swishing motion in the particles lining the starboard wall of the gap, like a wave in a curtain. Without warning, a large jagged body, at least several tens of kilometers across, parted the wall and came barreling toward me at about 10 kilometers per second—that's more than 30,000 kilometers per hour. This was only a third the speed of my orbit around the star, but because everything in my vicinity had been orbiting together, it was enormous compared to any of the speeds I had encountered lately. This body—let's call it an asteroid, it had that kind of shape and size—must have swung tightly around the back of another, much larger body somewhere beyond the gap and on the other side of the disk. Perhaps there was a planet in the outer reaches of this system that had already grown to the dimensions of an Earth, or even larger. Sudden, close-up encounters like this can act as slingshots. Two bodies, swinging close together because of a slight fluctuation in the disk's orbital regularity, suddenly find themselves pulled toward one another by immense, rapidly changing gravitational forces. More often than not, they are moving too fast to linger; instead, the lighter one shoots off at high speed on a new and reckless trajectory. These kinds of interactions can easily override the delicate balance established by the perennial but predictable tugs of a shepherd planet. I had been lulled into a false sense of security, thinking that the order imposed on the gap by my shepherd planet could not be disturbed by events occurring elsewhere in the disk.

The asteroid seemed to be heading straight toward me. I grabbed *Rocinante's* controls, planning to accelerate away once I determined whether the best escape route lay ahead of me, to the rear, or sideways. But before I could make a decision, the asteroid had crossed into the central gap, and I could see that it would pass well behind me. I instinctively relaxed, but I shouldn't have. The asteroid made a beeline for the shepherd planet (it was moving too fast for its path to have been curved noticeably by the planet's gravity) and then hit it squarely. A brilliant spot of light spread from the impact point. The patch just below the impact had been heated to nearly 3000 degrees and had vaporized. As the pressure under the impact point was released, a huge dome of vapor erupted, with a jet of luminous gas squirting outward from its highest point. Farther away from the impact, the planetary material had liquefied, and I could see globules of molten rock being thrown up from the surface. Within a few minutes the shock wave had spread halfway around the planet. No longer intense enough to melt the rock, it was cracking and buckling the planet's surface, sending shards of rock hurtling into space at high speed. A spray of these shards, mixed with flash-frozen molten globules, approached my craft, and I knew it was time to flee.

19

Finishing Touches

It was easier to outrun the spreading debris than to outpace my frustration, but I managed to do both. I was disappointed to find that a disk this far along in forming planets was still too dangerous a place for me to go poking around. I was henceforth to be relegated to a longer view of these systems, just as I had been banished from getting close to black-hole remnants of stars by the sickening effects of their tidal forces. This kind of frustration was beginning to seem dismayingly routine; I had no choice but to take my disappointment in stride.

The process of planet formation had proved to be much more gradual than I had anticipated. It was really a complex series of processes, with many steps. How different this was from star formation, in which the main body of the star came together all at once, leaving behind a disk that provided additional nourishment in what seemed a straightforward, two-step process. I recalled how difficult it had been, in the first planetary disk, to spot the evidence that tiny dust grains were starting to stick together and accumulate. Now, in the second disk, I found bodies that were almost but not quite the size of planets. I knew that they had to grow another 100- or 1000-fold in mass, yet the gaps that they had carved out around themselves had put the brakes on their rapid accumulation of additional matter. To go beyond this stage, the miniature planets had to latch on to a new mode of growth, and I had to decipher it without being able to explore the disk from within.

As I puzzled over how the protoplanets would get beyond the hiatus in their growth, I pulled far enough above the disk to survey several gaps at once. Some of the gaps were much wider than the one I had studied close-up, which indicated that their shepherd planets had greatly surpassed mine in mass. Errant asteroids, like the one that had slammed into my shepherd planet, must have etched the numerous faint trails that cut across the debris-filled zones between gaps. In some places, the gaps had wavy or indistinct walls, in contrast to the sharp boundaries I had considered to be the norm. I suddenly realized how naïve I had been to consider each gap, along with its shepherding planet, an isolated part of the disk. They all communicated, through the weak gravitational tugs that destabilized portions of the gaps (causing them to fill in) or changed the planets' orbits, or via the occasional asteroid hurled on a collision course from one to another of the shepherds' domains. The hope that I could find safety by attaching myself to a shepherd planet and accompanying it along its track had seduced me into assuming that each miniature proto-Earth would eventually grow into a full-sized version, while occupying much the same orbit it had been so zealously keeping clean. The gaps in the second disk had seemed like such nicely grooved tracks that I had thought the planets would stay put. In fact, the gaps, and the mini-planets that shepherded them, were ephemeral.

The asteroid-planet collision that had driven me from the second disk was my key to understanding the next stage of planetary growth. True, the particular collision that I had witnessed had been more destructive than constructive. On balance, the planet had lost a little mass in the encounter, but only because the asteroid had come in uncharacteristically fast and the planet was on the small side. The more common encounters would be less violent, and the planet would gain, on average. When the planet approached Earth size, even fast collisions like the one I had seen would be hard pressed to liberate a large fraction of the matter involved. At worst, shattered chunks of both colliding bodies would be thrown into orbit about the planet; most of the debris would rain back down on it, the remainder coagulating to form a satellite. Before I left Earth, many scientists had become convinced that our home planet had had at least one such collision, an impact with a Mars-size body (proto-Mars itself?) that had gouged out enough matter to form the Moon.

It seemed that protoplanets were susceptible to many possible fates. A lot of chance was involved. Sometimes a planet would grow mainly by repeated collisions with much smaller bodies. Asteroids might be thrown into the planet's path through the gradual disintegration of the gap walls. When this happened the collision would be slow, and the planet would absorb most of the asteroid's matter. In other cases the asteroid would be hurled into the planet's path, slingshot-style, following an encounter with another planet. The planets themselves also engaged in a very complex dance with its own internal logic. The planets' orbits would become locked into syncopations that were very hard to break once they formed. In such a "resonance," one planet might execute three orbits in exactly the same amount of time that it took a second planet to complete two. I remembered how hard it had been to learn to beat three notes against two when I studied piano as a child, but Nature seemed to do it effortlessly. A third planet might get locked to the first two by beating five orbits against three against two, a feat I had never even attempted on the piano. Four versus three versus two was also possible; the combinations were endless. Once locked in a resonance, planets could exert powerful influences on each other's orbits. Gaps and their attendant shepherds could drift, separate, or come together and merge. Sometimes, the planets encountered and even collided with one another. The motion didn't always look orderly. Smaller planets, especially, sometimes moved onto orbits that seemed chaotic. But the results of all these processes were greater variety, unexpected change, and, of course, the growth of planets toward their final dimensions.

Thus the final stage of planet growth was one of distant encounters, rhythmic orbital changes driven by the syncopated beat of the resonances, and bouts of chaos. The smaller planets gradually disappeared, their gaps merged into the larger bands cleared by bodies that were now truly becoming planet-size. The final growth itself was driven by much more violent collisions than I had hitherto encountered. This growth was not a clean process. Even as the larger bodies did their best to sweep up the finer particles of grit and dust, new debris was created continually as fast impacts nicked and ground down these bodies. Up until the very last stages—almost indefinitely, it seemed—planetary systems in formation were dusty construction sites.

The constant battering of the young planets led to gradual changes in their internal structures. As one impact after another flash-heated the planets' surfaces, their interiors began to warm. Brittle mineral amalgams became plastic and, in some locations, liquefied. Heavier materials sank to the planet's core, and the lighter minerals floated to the surface. This internal rearrangement was accelerated by dramatic deviations in the planets' trajectories. When a planet was knocked off its circular course by big impacts or by a strong gravitational tug from another planet, it could go into an oblong orbit around the star. In this case, the continuously varying pull of the star's gravity—the star's tidal force as the planet got nearer and then farther away—would also heat the planet's insides. It was comforting to know that a planet could get sick from being stretched by tides, just as I did near a black hole. Through the two effects, impacts and tides, the planets' interiors were cooked, and the undifferentiated conglomerates of dust, pebbles, and boulders evolved into the layered structure—core, mantle, crust—that we associate with home.

20

Virgin Worlds

I felt that I could now tell myself a story of how a planet like Earth might have formed, from start to finish. But I wanted to see whether the story was true. Even if I could not sail safely amid half-formed planets, I was determined to complete my study of planet formation from afar, if necessary. In a third system, only 4 light-years from the second, I hoped to catch the final phase of planetary gestation. Again I found the system first by spotting its dusty swath. This disk was even more open than the second one, and in fact the veil of dust and debris was so thin that you could see through it in most places. The dust was also glowing more warmly, in infrared light, which indicated that much of the starlight was getting through and warming it. There were fewer noticeable gaps and, on closer (but not too close!) inspection, fewer planet-size bodies. But each of the planets I saw was larger.

I could not resist playing time traveler, heading first for that zone, 150 million kilometers from the Sun-like star, where an incipient Earth might be preparing to host life. Here, the temperature could be just right to allow a planet to retain water and a thick atmosphere. Not that the atmosphere would be in place yet. The accumulation of water on the surface would have to wait for volcanic activity—an outcome of all that interior heating—or a rain of comets—sent into the inner reaches of the planetary system from its outer limits. If the chemistry went awry (if there was too much volcanic activity, for example), then the atmosphere might settle into a steaming sulfurous soup, in the manner of Venus. Too little volcanic activity, perhaps, and the planet might become a cold and arid world, like Mars, with most of its atmosphere evaporated or frozen. I was not expecting life, or even the setting for life, yet. But I was expecting to see a smooth, virgin world, ready for a completely new history to be written on it.

The planet was there, one of four or so, about the size of an Earth, but what I found was a brutally scarred body. This world was very close to spherical—it was so big, its gravity so strong, that any serious indentations would have filled in quickly—but its imperfections were just enough to remind me of a ball of clay shaped by hand, still bearing a child's thumbprints. Sheets of half-congealed lava, bounded by giant cracks and rifts, covered vast areas. The thick ring of debris that surrounded this planet, a mixture of jagged fragments and smooth globules, betrayed its recent history. This world had suffered a huge bombardment, perhaps of a magnitude similar to the impact that might have created the Earth's moon, and it had not healed yet. Had it been on the brink of hosting a flourishing ecosystem before its recent encounter? If so, how long before it would ready itself again? Or would some other fate intervene first?

That fate could lie in the hands of processes going on much farther away from the star. In the habitable zone around this star, a billiards game of planetary collisions was under way, but in the outer reaches of the disk, another game entirely was being played out. I had neglected the farther reaches of the other disks, partly out of impatience. Everything happened more slowly there, making it that much more difficult to see the processes of planet growth in action. I suppose that there was also an element of terrestrial chauvinism in my attitude, a bias toward regions and environments with Earth-like dimensions and conditions. Now that I understood the interdependence among vastly different regions of the disk, I knew this was an oversight I had to correct.

I headed outward, not perpendicularly away from the disk as when I had jetted away from the other disks, but this time skimming just above the surface. How different this scene was from the views I had enjoyed while skimming above the disk around a black hole! There the overwhelming impression had been one of luminescence and energy; here it was one of solidity and mass. I realized that this was mainly an illusion: The black hole had contained more mass than the star and planets combined, and I had been so much closer to the black hole that all aspects of the black hole's gravity had been vastly stronger. But just as we breathe an involuntary sigh of relief when setting down on firm land after a flight or a sea voyage, so I responded to a disk made out of rock and ice instead of superheated gases.

I haven't mentioned ice before because it was not a prominent constituent of the regions of disks I had been exploring. Ices—not just water ice but also those of ammonia and methane—had a pronounced tendency to condense on the surfaces of dust grains, but they could not do so if the grains were too warm. I had encountered ice-coated grains before, in the cooler regions of interstellar space, and I remembered likening some of these environments to snowstorms. Despite the ample shielding from starlight, ice could not condense on the grains that I had found in the inner regions of protoplanetary disks because too much warmth, mostly in the form of infrared rays, still managed to filter through.

But now, as I headed out to 3, 4, and more than 5 times the distance of the Earth from the Sun, conditions were growing positively chilly, and the dust grains were covered with rime. What was more striking was that the gas had not been winnowed from the dust at these distances. I had to keep adjusting my course to avoid flying through the disk, because the disk itself was flaring upward as the concentration of nimble light gases, mainly hydrogen and helium, first matched and then surpassed the concentration of solid dust.

While I was busily tracing the upward slope of the disk's gaseous atmosphere, I had a sudden sensation that the bottom had dropped out from beneath my feet. It was like that eerie feeling I once had on Earth when, having climbed steeply to avoid the turbulence in a thunderstorm, I cleared the edge of the anvil and suddenly looked down on a flat cloud deck tens of thousands of feet below. My craft was flying in trim, then as now, but it was impossible to avoid the momentary sensation of falling off a cliff. I looked toward where I thought the disk should be and found that it was gone. I was in the biggest gap I had ever seen, so big that I couldn't make out the other side. If it had been cleared by the gravitational effects of a shepherd, this had to be one enormous planet.

This time, it did not take long to find the shepherd planet. Although still much smaller than the gap it had cleared, this substantial body was shining brilliantly with its highly reflective clouds. I estimated it to be 3 times heavier than the planet Jupiter—making it about 1000 times heavier than Earth—and of similar architecture. Its rocky core would have been perhaps 10 times heavier than Earth, perhaps a bit more. But the vastness of its bulk would have been contained in its deep, dense, gaseous atmosphere. In these outer reaches of the disk, where the gas had not yet been separated from the dust, there was 100 times more raw material to incorporate into the body of a planet. This proto-Jupiter had greedily sucked up the gas. It had sucked up the grains, too, complete with their icy coatings. Stripped from their grains, the ices formed crystalline suspensions of ammonia and methane as clouds floating in the planet's envelope; these are what made the planet shine so brightly.

This proto-Jupiter had not yet finished growing. Unlike the young Earth-like planet that I had just visited, which had relied on the complex dance of orbits and random collisions to throw raw materials in its way (and was paying the price in its tortured and scarred visage), this monster planet was skimming a steady supply of gas off the outer wall of its gap. Gas, unlike pebbles and boulders, was not so easily herded by gravitational forces alone. Its pressure was perpetually leaning on it to fill in the gaps, squeezing it toward any vacuum. And when proto-Jupiter passed by in its orbit, the wall of gas was unable to resist the temptation. A gaseous wave erupted from the wall of the gap and licked out toward the planet, pulled by the latter's gravity. I followed along in the planet's orbital wake, admiring the swirling eddies that were excited by the passing behemoth, along with the graceful stream of gas that encircled the planet and disappeared into its thick atmosphere.

Moving farther outward in the disk, I found three more giant gaseous planets. They lorded over much larger domains than their Earth-size counterparts but were smaller than the proto-Jupiter I had just visited. Each one was still growing by stripping gas off the walls of its gap, but at a much lower rate than proto-Jupiter. The inability of the gaps to stay completely clear of their shepherding planets made me realize that there was yet another dynamic at work here, one that could endanger the Earth-like planets closer to the star. When each giant planet disturbed the walls of its gap, the interaction was not entirely one-sided. Newton taught us that every action elicits a reaction; thus the pull of a proto-Jupiter on the gas and dust that surrounded it would, in turn, affect the planet's orbit. In the situations I had observed, each of the planets was dragging the gas along with it, speeding the movement of the gas around the star. Newton's law implied that the gas would pull back on the planet, with equal and opposite force, thus slowing it down a bit in its orbit. As a result these enormous planets were spiraling, ever so gradually, toward the star. I knew immediately that this spelled danger for proto-Earth and its companions. Already, the gravitational tugs of these distant but massive worlds could be measured in the subtle motions of asteroids close to the Earth-like planets, and even in the motion of proto-Earth itself. Once these giant bodies got much closer, the effects would be drastic. If they didn't come so close to proto-Earth as to send it careening out of the planetary system altogether, they might fling a fatal barrage of asteroids or planets into its path or destabilize its orbit so that it drifted into the star and vaporized. There was no telling when or if proto-Jupiter or the other giants would tire of their inward march.

And there was no predicting what would become of the giants themselves. Would they march inward in lockstep until they plummeted into the star? Would they spread themselves out in a stable configuration that would leave them intact, whether or not there was room for a few Earth-like planets as well? Or would they lose the ability to coexist, invade one another's territory, and fight a gravitational duel to the death of all of them but one?

Given the extreme uncertainty of the planets' fates, I realized it would be prudent not to invest too much emotional energy in any particular planetary system. I therefore left battered proto-Earth to its future, whatever that might be, and withdrew from the third disk I had investigated. After visiting these three systems, I continued to explore the planetary systems of Orion, but in a more desultory fashion. I seemed to find planets nearly everywhere I looked. Even the fully formed planetary systems, when spied from afar, showed the signatures of dust; soon I learned that this was a quick way to spot the most likely candidates. I remembered that even the Sun's system had retained a weak, dusty signature after all its billions of years. It was no surprise, therefore, that the younger systems of Orion—tens of millions of years old, or less—should show dust, and the visible signs of larger debris, in abundance.

I pondered the visible indications of our planetary system's history. In the Solar System there was, first, the zodiacal light, a shaft of sunlight scattered by interplanetary debris, that shot above the horizon for a couple of hours before twilight began and for a similar period after it was over. People who lived under the darkest skies (such as myself and my fellow stargazing Montanans) used to boast that this faint glow—tracing the ecliptic, the path of the planets on the sky—interfered with their view of the stars. But this material was not exactly a fossil of the era when planets were little more than consolidations of dust. More likely, it was relatively recent debris, thrown into orbit from the occasional collisions of the last few hundreds of millions of years. Second, there were the belts and clouds of debris in the outer Solar System: Kuiper's belt and Oort's cloud, extending far outside the orbit of any planet, swarms of icy bodies that had presumably been ejected from the inner planetary system by the gravitational action of Jupiter, as though by a slingshot. They really were fossils of the forming Solar System. These swarms were very important to the denizens of Earth, because they were the sources of comets, whose appearances sometimes had changed the course of history, and whose far less frequent collisions with Earth had surely changed the course of evolution—for example, wiping out the dinosaurs. But not that much of the primordial rubble was left, and what did remain was spread so thin that it would have been difficult to detect from a distance. An incredibly thin tissue connected our Solar System to its origins. But the younger planetary systems wore their aura of dust for hundreds of millions of years.

Seeking snapshots of the final phase of planet formation, I toured one young system after another in Orion's vicinity. Each planetary system I chose to visit encircled a star not unlike the Sun, but that merely reflects my sentimental side. Most kinds of stars—with the exception of the hottest, most massive stars and those with a close stellar companion—had planets. Binary stars tend not to have planets, presumably because the gravitational attraction of the companion disrupted planetary orbits or perhaps prevented the dusty disk from settling down. Every system I visited had a unique arrangement of planets, and few of them resembled the Solar System. There were planetary systems in which giant gaseous planets orbited the star close-in, taking only a few days to circle the star. Had these planets spiraled in from much farther away? And if they had, were they about to commit suicide in the star's furnace, or would something intervene to halt their plunge? Or had the disks in these systems retained their gas much closer to the star, somehow allowing the planets to condense at this small radius?

Other systems had planets of various sizes on orbits that swung widely from maximal to minimal distance. Such orbits were probably acquired after the planets had condensed, perhaps through some catastrophic encounter, and they were not conducive to life. Still other systems contained a wide array of small planets in haphazard orbits. These systems clearly could have used a Jupiter to police them. As it was, they never outgrew the era of devastating collisions, and their planetary surfaces never had a chance to settle down.

I gradually grew weary of this cavalcade of planets. I couldn't see a pattern; rather, I should say that I couldn't see a pattern that appealed to my anthropomorphic sensibilities. Planets obviously formed easily, provided the raw materials were available. It was astonishing that these tiny particles of dust could cement themselves together, one by one, until they had formed a pebble, a boulder, a mountain, finally an Earth. But it was far from certain that planets would form in the right place, with the right mass and composition to create and foster life. There seemed to be no imperative driving the debris left over from the formation of a star to organize itself according to the blueprints of our well-ordered Solar System. Far from it: If I had to guess on the basis of my travels, I would venture that our planetary system's structure—the relative stability of its Earth-like planets, in particular—was a fluke. The most one could say was that planetary systems were so universal, and so diverse, that some of them were bound to present the fortuitous conditions favorable to life. How easily life would form once such conditions existed . . . that topic would have to wait for some other exploration.

I looked back over the trajectory of my journey to this point. It seemed a good time to take stock. I had moved from the most exotic places in the Galaxy—black holes, and eerie neutron stars—to about as close to home as I was likely to get without actually returning home. Yet the diversity of the phenomena I had seen did not strike me as the defining quality of this voyage. What struck me was the unity of what I had seen, and at the same time its complexity. Gravity—a simple force, purely attractive—was capable of creating this enormous diversity of motions and structures. Jets, disks, stars, planets. Black holes with disks and jets, stars with disks and jets. After all that, how could

I be astonished at the diversity of planetary systems? What really distinguished planets was not the process by which they formed (this was just another example of the routine operation of gravity) but the stuff they were made of.

The mighty dust grain. Why were these minerals made of heavy elements—silicon, carbon, oxygen—here at all, and why were they in solid form? This seemingly mundane question, focused on some tiny objects I had regarded as nuisances throughout most of my voyage, suddenly assumed paramount importance. To answer it, I would have to visit some other kinds of objects I had taken for granted.

Part Five

Evolution

21

The Blob

Will you excuse my naïveté in believing that stars should be spherical? As a theoretician, what else could I have believed? Because the force of gravity pulls equally in all directions, any star should arrange itself into as perfect a globule as the Sun. There would be no reason for it to stick out more in one direction than in another. Furthermore, the gas that composes a star is extremely fluid. One could not so much as pile up a respectable mountain or dig a deep canyon on a star, and have it last, without heroic efforts at maintenance. Any deviations from a spherical shape would be quickly washed away. If a star were spinning rapidly enough, then I could understand how it might bulge about the equator, the result of centrifugal force flattening it slightly. Jupiter, which spins so fast that its day is only 10 hours, presents a noticeably oblate profile as viewed from Earth through any small telescope. Yet even that whirling dervish of a planet is pretty close to spherical.

But not this star. *Rocinante* was hovering above a vast, heaving plain that filled more than a hemisphere of my vision. I was trying not to get too close for comfort. Although astronomers classified this as a "cool" star, at a temperature of 3300 degrees (little more than half that of the Sun) its surface was hot enough, especially given that this heat was emanating from such a vast area.

All told, 200,000 times more power than the Sun emitted was streaming into space through the luminous smog that spread out beneath my craft. What passed for a surface—as far as I could tell—enveloped a body with a diameter about 30 percent larger than Jupiter's orbit around the Sun. This was a bit hard to take, as attached as I had lately become to the conceit that planets like Earth might exist nearly anywhere. A planet, eking out its existence on an Earth-size orbit, would be submerged beneath more than 80 percent of this star and would have vaporized long ago. I immediately checked my thoughts, realizing that I was begging an important question. If an Earth-like planet had ever had the opportunity to orbit this star, I must now be witnessing the bloated descendant of what had originally resembled a normal star, perhaps like the Sun. I shuddered at the fate that must have befallen any denizens of that planet. Or had the star always been like this? I was here to find out.

I thought I knew the answer, because this star had been an object of intense study among my colleagues. The famous star Betelgeuse was the garnet that pinned the cloak to Orion's left shoulder, as viewed from my childhood home. I had voluntarily clipped my wings at this stage in the journey. Although it was not a part of the Orion "star formation" empire, Betelgeuse was near enough to it. Indeed, for a traveler lately departed from the Orion Nebula, it was "on the way back" toward Earth, a mere 500 light-years away from home, only one-third as distant as that marvelous nebula.

Was it a failure of imagination, fatigue, or the draw of home that led me even closer to Earth on this next leg of my journey? Perhaps some of each, but it was also the openness of this star's architecture. This red giant—rather, I should say, this body that my colleagues classified as a red *supergiant*—seemed ready to burst, a loose bag barely able to conceal what was inside. The secrets inside the star were what I sought. All other stars I had seen, even the ones just condensing, had seemed so tightly composed that all hope of probing the interior seemed vain. Here I thought I had a chance, although I was soon to learn how circumscribed that chance was.

The carbon, the oxygen, the silicon, the calcium, the iron—all the elements that make dust and rocks, giving planets their rigidity, had to be created inside stars. The interiors of stars were nuclear pressure cookers where a kind of permissible alchemy took place: the transmutation of hydrogen and helium into heavier elements. The pristine Universe, without stars, would have consisted solely of those two simplest of elements, salted with a dash of the next element in the hierarchy of atomic weights, lithium. But in which stars, and when in the life of each star, did the more complex raw materials form? And how, once formed, were they dispersed into open space, to become the planetary bodies I had lately encountered?

My attention was drawn back toward the extraordinary light that bathed my craft. In keeping with the temperature of the star, this light was intensely red, so much so that it gave everything in the cockpit an eerie red glow except the blue materials, which were turned jet black. To say that there was something unsettling about this star would be an understatement. I was hesitant about describing its "surface" for very good reason: There was no crisp limb, such as one associates with the Sun and every other star I had seen. Once my eyes became accustomed to the light, however, I could perceive a structure to the star's upper layers. There was an opaque "surface" after all, but it was indistinct and submerged beneath a translucent zone, glowing the same deep red but gradually shading to transparent.

As I watched, both the translucent layer and its opaque undercoat heaved and buckled. None of these bumps and knobs were permanent features. They swished from side to side, and up and down, slightly changing in color and brightness as they moved. Then I noticed the feature that must have made their activity so disconcerting. As the translucent layer heaved upward, toward me, the opaque floor seemed to drop away. The upper layers of the star darkened and cooled, and I had the uncomfortable sense of being suspended above a void.

The darkened area grew to cover a good quarter of the star's surface and would not stay still even then. It began migrating around the star, so slowly and deliberately that it would take years to make a circuit. Moving out from under my vantage point, it was replaced by its opposite, as the opaque floor welled up, glowing brighter and protruding into the translucent layer, narrowing it almost to the point of invisibility.

As I stared at these vast swells I began to grow seasick. What could make the star slosh so vigorously? Wasn't there supposed to be equilibrium between the inward pull of gravity and the outward push of the star's interior heat, manifested as pressure? If so, then why all this motion, as though the star somehow could not find its favorite place of repose? I had sorted through the concept of equilibrium—or so I had thought—in its most challenged state, when I had visited the neutron star at the center of the new Crab Nebula. There the atoms had been crushed to the point where the electrons and protons had merged. At the last instant, the degeneracy effect and the nuclear force of repulsion among the remnant neutrons had saved the star from collapsing to a black hole. Surely, this star's much weaker gravity could manage to preserve a balance with the ordinary, less violent pressure of heated gas. Yet the equilibrium near the surface of Betelgeuse seemed to exist only in a loose sense. It was an equilibrium that permitted—or required—enormous waves of matter to sweep across the star, allowed frightening pulsations to erupt, and kept one guessing where the star ended and space began.

I began to doubt whether there was equilibrium at all, or even a boundary to this star. No matter how far I drew back from the star, I never seemed to reach the sparse conditions of interstellar space. It seemed that this star's atmosphere went on forever. Was it disassembling itself without my realizing it? I had earlier encountered winds from stars. Even the Sun had a puny wind, and the hot, massive stars of Orion belted their surroundings with powerful gales. As I moved about I detected no such motion. Only when I drew to a halt, trying to position *Rocinante* as steadily as possible with respect to the center of Betelgeuse (insofar as I could determine where that center was), did I detect the outward flow of matter leaving the surface. The wind was slow—a breeze, really, the gas moving only a few kilometers every second, compared to hundreds of kilometers per second for the Sun's wind and thousands for the winds of Orion. But Betelgeuse's wind was dense, and I estimated that more than a Sun's worth of matter would leave this star within 100,000 years.

Surely the loss of so much matter so quickly would have a devastating impact on a star. But this body was so huge that it was hard to believe it lacked almost infinite stores of matter. I maneuvered *Rocinante* into orbit around the giant to deduce its mass and received a shock. It was only 20 times the mass of the Sun: At its rate of evaporation, this star would be gone in little more than 1 million years. My shock was short-lived, though. The Sun, I knew, had started out with enough nuclear fuel to last for 10 billion years. It was not so heavy as Betelgeuse, or as the stars of the Trapezium, but it was parsimonious, using up its store of hydrogen slowly. In keeping with its stinginess, the Sun's wind was so weak that it would have little impact on its development, even over all those billions of years. Betelgeuse and Orion's hot stars were profligate, burning their nuclear candles at both ends. Yes, they had several times the Sun's mass, but they burned their fuel supplies thousands of times more quickly. Even without winds, they could not last more than a few million years. I recalled that the winds from the Trapezium stars had also carried away large amounts of matter, perhaps even as much as Betelgeuse was dispensing. It was curious that potent winds seemed to go along with other signs of prodigality.

Betelgeuse and Orion's Trapezium stars parted company in at least one important respect. The Trapezium stars were relatively small, with diameters only a few times larger than that of the Sun. They shared the compactness, sharp surface, and concentrated gravitational pull that I had learned to associate with stars. Betelgeuse had a similar amount of matter, but it came in a much larger package. The star was grossly distended, occupying a volume over 2 billion times that of the Sun, and its gravity, spread over such a large distance, was so weak that its interior could not be nearly so hot as that of the Sun—let alone that of a massive hot star like one of the stars in the Trapezium. If it became that hot, it would blow itself apart in a year!

If only I could dip a thermometer deep into this churning blob, but diffuse as it was compared to other stars, the heat coming off its surface and its opaque screen of red-glowing gas kept me at bay. Most of the way to the center, the temperature could not have risen much above 100,000 degrees. That would not be enough for nuclear reactions to create the amount of light that was bathing *Rocinante* and shining into space in all other directions. Thermonuclear reactions need high temperatures, to bang the atomic nuclei together with sufficient force, and high density, to ensure the fierce collisions occur frequently enough. Betelgeuse had neither, at least within its huge mottled envelope. Deep inside this star, however, there had to be someplace where nuclear reactions did occur, and they had to be sustained at a rate that far outstripped stars like the Sun. A prodigious amount of energy was forcing its way through the star so insistently that it kept the star's outer layers off guard, unable to settle into a passive role as the mere conveyor of stellar luminosity. The envelope of Betelgeuse was actively involved in transporting its energy, a role that rendered it restless.

The star took another big gulp, and a new abyss seemed to widen under my craft. I half-expected the curtains to part and even this nebulous but opaque surface to give way fully, allowing me to see down deep into the star's insides. Then the displaced layers of gas rushed back in a tsunami that swept closer to the underside of *Rocinante* than I could stomach. I pulled farther away from the star.

22

Divining the Interior

The delicate imbalances that characterized the surface of Betelgeuse made it seem fragile, but something about the star hinted at a concealed ferocity that I did not wish to challenge, even if I could. I was not yet ready to depart from the vicinity of Betelgeuse, but it was apparent that I would have to fall back on a much greater dose of theory than I had anticipated. Of course, I had studied red giants and supergiants ages ago. The theory of stellar structure had been *de rigeur* for students of astronomy. It was the crowning glory of theoretical astrophysics in the mid-twentieth century; the prediction that stars should grow to the enormous dimensions of giants late in life was one of its greatest triumphs. From the remote point of view of an astronomer on Earth, it would have been thrilling to know everything about a certain type of star without having seen one—and the theory was good enough that this was almost possible. But the features of Betelgeuse that struck me most viscerally were those where the theory was weakest. Theoretical sketches of a smoothly distended envelope paled beside the great waves of matter that sloshed around the star as its internal supply of energy struggled to get out. Statistical descriptions of the bubbling and turmoil that went on just below the surface had never really captured the sudden emergence of turbulent cells that I saw growing to cover degrees of longitude. And the theoreticians had never quite been able to predict just how quickly such a star would erode, mostly in spurts and gasps, of its own volition.

Thus it was with some frustration that I had to recall my rusty theoretical tools and infer the hidden operations of this star, trying wherever possible to enrich my deductions with the sensations afforded by my presence so close to this body. Perhaps it was a fitting irony that what I understood of Betelgeuse, close up, would have to proceed from theory rather than being gleaned via proximity to the beast. Even the space traveler, who could explore a planetary surface or atmosphere with relative ease—perhaps with a robotic probe, but directly nonetheless—was still excluded from direct experience of stellar interiors.

Astronomers had once thought that stars shone by squeezing the heat out of themselves. The squeezing came from gravity, and this nineteenth-century idea foreshadowed the later discoveries that objects with really strong gravitational fields, such as black holes and neutron stars, could become luminous by sucking in matter. But this would not work for ordinary stars, because their gravity was not strong enough. The Sun, if powered by gravitational squeezing alone, would last only 30 million years. Thus the discovery of thermonuclear fusion, the heat-yielding reactions that combine smaller atomic nuclei into bigger ones, must have come as a revelation to my academic forebears. There was a short hiatus in which everything stellar was thought to derive from nuclear power, but then gravity returned in a subtler role. If gravity did not power stars directly, its inexorable pull, combined with the changing chemical makeup of the stars as they burned away their supplies of fuel, made their internal structures, and appearances, change with time. Gravity made stars age.

I recalled the outline of how stars evolve. All stars start out with a core rich in hydrogen and a temperature just high enough to fuse that hydrogen into helium at a moderate rate. More massive stars use up their hydrogen much more quickly than less massive stars; this is why they are so much more luminous and burn themselves out so much more quickly.

At first the temperature is highly regulated and differs little from one star to another. In fact, the center of a young star possesses one of the most elegant thermostats known. If the temperature ever climbs slightly too high, the nuclear reactions run haywire and push the core apart, cooling it and quenching the reactions. If the temperature is too low, then the reactions—extraordinarily sensitive to heat—effectively shut down, allowing the core to contract under its own weight until the thermometer rises and nuclear reactions resume.

All of this proceeds smoothly, as long as hydrogen is evenly distributed throughout the core. But this situation cannot last indefinitely, because stars incinerate themselves from the inside out. Eventually hydrogen becomes scarce in the center of the core. The helium that is left behind may burn later on but for now the temperature is not high enough. The nuclear furnace retreats to a shell at the core's margins, stuck between the star's envelope, where it is too cold, and the star's center, which is starved for fuel. Because the center of the star is no longer burning, the thermostat fails, and there is nothing to stop the core from shrinking. The burning layer—the shell—shrinks along with it.

These are the conditions that set the stage for a red giant. Once nuclear reactions begin to run out of fuel, what's left of the furnace gets pulled more tightly together by gravity. As it is compressed, it gets hotter and burns all the faster, pumping out ever-increasing amounts of heat and light. This was the first conundrum of stellar aging: As stars run out of fuel, they grow *brighter.* Energy is forced into the outer layers of the star faster than those layers can handle it. The heat gets trapped, so the envelope expands, its surface ballooning so enormously that it actually cools down even as it is pumping out more luminosity. Hence the second conundrum: As stars grow brighter, they grow cooler.

As I recapitulated these classic theoretical arguments, I began to understand the basis for my misgivings about the benignity of Betelgeuse's deep interior. The soft and fluffy envelope concealed a searing core, shrunk down to near-Earth dimensions—a hundred thousand times smaller than the star! The core was a world unto itself, so self-contained and compact that it cared little what the envelope was doing. Thus the outcome of stellar aging, the legacy of gravity, was a gradual disconnection between the star's core and its envelope.

No wonder the stability of the envelope seemed so precarious. The shroud surrounding Betelgeuse had been pushed to the limit, inflated to the point where it had cooled down to 3300 degrees, about as cool as a star's envelope could get. There was a reason why red giants and supergiants could never get cooler than this. It was, perhaps, the third conundrum of red gianthood: If a star got too cool, it would release too much of its energy all at once. This bizarre effect was the result of chemical behavior that began to occur at such relatively low temperatures. It was as though the outer layers of the star were a window made of a strange substance whose transparency depended on how hot it was. If the temperature got too low, the window would become more transparent and lose its insulating powers, allowing the stored heat to stream out faster. The rapid loss of energy would cause the envelope to shrink, but—curiously—a shrinking envelope grew hotter, losing its transparency and beginning to store up heat once again. In this way, the temperature would bounce around the 3300 degree threshold, not growing markedly colder but not remaining very stable, either. This, more than anything, was why I saw the opaque "surface" of the star heaving up and down. What appeared to be vast vertical motions were only partially that. I was also seeing more or less deeply into the star as its transparency fluctuated wildly.

23

Nuclear Alchemy

As though to underscore Betelgeuse's unpredictability, another swell of luminous garnet fluid converged beneath my craft and surged into the space around me, cooling and darkening as it expanded. This gust of matter, liberated from the star, reminded me of my original motivation in visiting Betelgeuse: to trace the raw materials of planets back toward their roots. I had not succeeded yet. The "red giant story," as far as I had recounted it, was not useful in producing the elements heavier than helium. The helium at the center of a red giant was an inert mass not hot enough to fuse into anything heavier; the energy source of a red giant was still provided by the conversion of hydrogen—in a shell surrounding the core—into yet more helium.

Who needs stars to manufacture helium? About 1 in 13 atoms in the Universe—1 part in 4 by mass, because each helium atom weighs as much as 4 hydrogens—would have been helium to start with, just after the Big Bang and before any stars had formed. And helium, hardly reactive and nearly always gaseous, is useless for forming solid planets.

But Betelgeuse was creating more than helium. It was no red giant. It was too big, too luminous for that—it was, indeed, a supergiant. I needed to follow the star's interior saga one episode further. The red giant story fit Betelgeuse in outline but not in detail. The story so far had introduced, elegantly, the character of the bloated envelope, its expansion, why it was red, and the growing disengagement between the assertive core and the passive shroud. All of these features would carry on to the higher level of gianthood, with a vengeance.

Once again gravity is the culprit. As hydrogen burns all around it and helium accumulates, the core continues to shrink and grow hotter. Eventually, it becomes hot enough for helium to fuse into carbon, and then (if the star is heavy enough, as Betelgeuse was) for carbon to fuse into oxygen. At first this would happen only in the center of the core, where the helium is spread evenly. The even burning of helium would create a new thermostat effect, like the one that prevented ordinary stars from turning into giants. The red giant would shrink and would come to resemble an ordinary star once again, only brighter and hotter.

However, the brush with normalcy would be short-lived. Helium at the center of the core is quickly used up—in only a few hundred thousand years, for a star like Betelgeuse—and the nuclear reactions again retreat into shells surrounding the core. But this time the multiple layers of the nuclear inferno are burning so ferociously that the star grows to supergiant size and luminosity.

Thus it would seem that Betelgeuse had already produced some of the materials out of which one could mold the Earth and the other planets I had visited. According to the theory, carbon and oxygen would be joined by some other trace elements. But would they be disseminated into space? Could Betelgeuse's wind be the medium of their dispersal? My measurement of the wind's composition, as a fresh gust blew past, seemed to bear out the idea. The wind was cool and dense enough for molecules to form, even for solid grains to condense, and condense they did. There were the common terrestrial gases carbon monoxide and carbon dioxide, some odd rarities such as the oxides of zirconium and titanium, and soot—the carbon dust that could someday fill a protoplanetary disk. The disconnection between core and envelope evidently had not been so absolute as it had seemed. The churning streams of matter carrying energy to the surface had mixed with some of the freshly cooked elements and swept them aloft.

But something told me that I still didn't have the full story. As I puzzled through what must be going on in the hidden interior of Betelgeuse, I was forced to accept that Betelgeuse had once been a massive, hot star, perhaps identical to one of the stars in the Trapezium. Such stars are rare, and although each one carries more mass than an average star, they could not, all together, contrive to produce all the carbon of the Universe. Moreover, they do not release into their surroundings all the carbon they do produce. For a star like Betelgeuse, life as a red supergiant, spectacular as that stage may appear, would not be the climax. Its core would go on getting hotter and hotter, its evolution accelerating as it produced successively heavier elements until the center of its core was made of nothing but iron. Before its slow, dense wind could have completed the dispersal of whatever lighter elements are left, its core would have done something much more dramatic. The stage at which it produced and dispersed elements like carbon would be just a milepost along its road to something grander, not the ultimate use of its talents.

Accordingly, it must be the more widespread types of stars, those heavy and old enough to have exhausted their nuclear fuel but not much heavier than the Sun, that spread around the elements such as carbon. Perhaps I should have visited one of them. They would also go beyond the red giant phase to become supergiants and to create carbon, even turning some of the latter into nitrogen in an elaborate nuclear barn dance outside their inert cores. Would they, too, mix their newly formed elements with the unfused gas of their envelopes, dredging the enriched gas up to the surface and releasing it in a wind?

It sounded like an awfully cumbersome process. It would be much more helpful, I thought, if a star revealed itself in a more systematic way. Peeling away layer upon layer, it could offer up the results of its nuclear alchemy, the strata of newly cooked elements, in order. If they had been somehow mixed, the cauldron stirred to ensure that the stew wouldn't be lumpy, then the star's disassembly would reveal that, too. In fact, some stars had to give up not just their outer layers, but also much of their deep interiors, if deep space were to receive enough raw material for the building of planets and, for that matter, new stars.

All this time I had been allowing *Rocinante* to drift away from Betelgeuse. From this distance I could finally take it in as a whole: the way it barely seemed to cohere, the sloshing that threatened to dismantle it at any moment, the unsteady wind that was gradually eroding it. I was far enough away to grasp the crudely spherical but uneven shape of an entire hemisphere. As I watched, it erupted again. A wave seemed to envelop the whole star, it pulsed, and a shell of gas flew off into space. Had I had the time, I would have been tempted to wait for the dénouement: the loss of this star's entire envelope and the final revelation of what was inside.

It's lucky I didn't. I already suspected that this star's fate would be dangerously violent, as I was to appreciate—in an encounter with a different star—later in my journey. Betelgeuse would not suffer a dénouement but rather an apotheosis of self-destruction. But there were plenty of other stars that had dismantled themselves, benignly and recently enough that I could perform just the study I proposed.

24

The Dumbbell

All amateur astronomers are familiar with planetary nebulae. Just the name conjures up an amalgam of the two most irresistible targets for homespun telescopes. The discoverer of the class, William Herschel, had named them deliberately, only four years after he had discovered a real planet (the first such discovery since ancient times), Uranus, in 1781. These nebulae often presented smooth, bright disks, sometimes so compact that their shapes were difficult to make out in telescopes of his time. They can be bright and glow with an even fluorescence that could well be mistaken for the reflection of stellar light by a planet—not that Herschel ever took them for solid bodies. They also show more complex structures. The first one identified by Herschel possesses two luminous extensions, protruding on opposite sides, which led observers to refer to it as the "Saturn Nebula." Another, a staple of my days as an amateur stargazer, was known as the "Ring" and famed for its delicate elliptical shape and apparent hole in the middle.

I was headed toward one of the most famous planetaries, known colloquially as "The Dumbbell." This was one of the largest and least well defined of the planetary nebulae, a full 2 light-years across. The name was apt only insofar as the nebula did not approximate a full disk or ellipse but seemed incomplete, with large bites taken out of opposite sides. This gave it a linear or box-like appearance; in three dimensions one could imagine it as a narrow-waisted hourglass. At the ends of the hourglass, intact arcs of the remnant circle possessed bright rims, and one could make out—or imagine that one did—a ridge of luminosity connecting the arcs through the middle of the ruined disk. One can only suppose that, to some nineteenth-century astronomer with an imperfect lens, the nebula must have resembled a pair of weights hanging off the ends of a bar.

As viewed from Earth, the Dumbbell lay in the direction of the constellation Vulpecula, the fox, nearly halfway round the sky from Betelgeuse and nearly twice as far from home, 800 or 900 light-years. I took a direct route, which meant that I passed closer to Earth than I had been since leaving—100,000 years earlier, by Earth time!—but in my haste to follow through on my quest, I did not stop. What this route did afford me was a view of the nebula not too different from the one I had known as a child.

Planetary nebulae are justly acclaimed for their great range of fluorescent colors, often arrayed in beautiful and orderly patterns. The extraordinary diversity of colors is no mystery. Just as the hot stars of the Trapezium illuminate the Orion Nebula, each planetary nebula possesses its own illuminating star. But the stars that light up planetary nebulae are hotter than those of ordinary nebulae. The powerful ultraviolet rays that they emit tear into the atoms of the nebula with greater destructive impact. Whereas the light from the Trapezium stars can knock one or two electrons off the oxygen atoms of Orion, fry most of the hydrogen, and wreak mild havoc on atoms of other chemical elements such as neon, sulfur, iron, and helium, it cannot, for example, tear both electrons off helium atoms simultaneously. But in a planetary nebula this is done with ease. The result is that the diverse opportunities for atoms, ions, and electrons to recombine in different permutations yield a much richer stew of atomic activity and hence a richer palette of colors. The gas in a planetary nebula is also hotter, fostering additional fluorescence as the atoms knock into one another with greater force.

There is another, more fundamental difference between the Orion Nebula and a planetary nebula like the Dumbbell. Orion is a patch of raw material from which new stars and planets are condensing—matter derived from a molecular cloud, which had in turn scavenged it from interstellar space. A planetary nebula is at the opposite end of the recycling process. It consists of matter that is being returned to interstellar space. The Dumbbell is precisely the substance of an old star's interior, the insides of a defunct red supergiant, expanded and made visible. The Trapezium cluster consists entirely of newborn stars; the star at the center of the Dumbbell is on its deathbed.

As I approached the nebula, I decided to maneuver my craft so as to enter along the axis of the "dumbbell." The comparatively smooth distribution of interstellar matter in this locale gave way to a lumpier texture even before I reached the bright rim that marked the threshold of the glowing gas. I surmised that I had already crossed into the region that had been overrun by the wind from the red supergiant. Closer in, I began to perceive the glow of gas being attacked by the vanguard of ultraviolet rays, first in a spotty pattern and then more uniformly. The atoms were not being treated too roughly here, a fact I attributed to the relative mildness of the photons that had managed to penetrate this far. All of the most extreme ultraviolet rays from the star had been absorbed—used up—by the gas closer in and were not reaching this distant outpost of the nebula. Consequently, the dominant colors were those of the atoms and ions that were easiest to knock apart: the reds of hydrogen and of weakly disturbed nitrogen, for example. I was slightly surprised that there was no sensation, other than visual, as I crossed into the outermost luminous arc. I had half-expected a slight bump, conditioned as I had become to associating sharp and bright boundaries with shock waves. But this gas was still more or less the undisturbed effluence from the old supergiant. The flow here was leisurely. As I knew from my experience at Betelgeuse, the wind coming off the extremities of a red supergiant might well have had velocities less than the 20 or 25 kilometers per second I measured here. The wind's speed also gave away the time that had elapsed—30,000 years—since this gas had taken leave of its parent star and joined the ritual of unraveling that I now witnessed in its latter stages.

I progressed toward the star. More of the intense ultraviolet radiation was able to reach my craft as I put layer after layer of the nebula behind me. The reds of the hydrogen and weakly ionized nitrogen gave way to the green of doubly ionized oxygen (a signature of hot nebulae everywhere) and then to the blue-green of completely demolished helium, that trademark of planetary nebulae. I sought evidence for another trademark of planetary nebulae, the sharp spatial demarcations that often layered the different levels of ionization. In the Ring Nebula, for example, photos taken through blue filters showed that the "finger hole" of the ring, which so beautifully frames its illuminating star, is actually filled with the faint light of doubly ionized helium, whereas the ring itself glows brightly in green oxygen, wound round with red filaments. The idea was that the ultraviolet rays grow gradually weaker, and softer, with distance as they propagate through the nebula. This effect was supposed to be aided by the progression of winds emanating from the central, dying star: first the red supergiant wind, slow and dense, then the gusts getting faster and faster as the star unburdens itself of its shroud and expels matter straight from the nuclear burning layers. The faster winds would plow into the supergiant's slow breeze, sweeping it into a shell that, in the case of the Ring Nebula and some other famous examples, appears as an annulus on the sky. Inside the shell, the bubble of tenuous gas left over from the fast wind would provide unimpeded access for the harshest ultraviolet rays and thus become the site of battered helium's bluish glow.

Many planetary nebulae possessed even more complex and dramatic structures. The protrusions—handles, or "ansae"—that graced famous planetaries such as the Saturn Nebula were thought to be narrow jets of matter spurting in opposite directions from the central star. It would be a nice symmetry if brightly glowing jets ushered out a dying star much as I had seen them usher infant stars into the Universe in the Orion Nebula. But I could not perfect the analogy. The young stars had their accretion disks, whose swirling motions seemed to be connected to the creation of jets in such diverse circumstances as protostars and the X-ray binary SS 433. In a planetary nebula there was no disk. Some astronomers speculated that the star had created a chimney within itself, perhaps along its rotation axis, or corralled and focused by the motions and gravitational tugs of an unseen binary companion. The jets would then be streams of matter propelled out through the chimney.

I was never able to determine what caused the elongated symmetry of the Dumbbell Nebula. Any old jet tracks that persisted here were intermittent and indistinct. Had the ejection occurred through a broad cone, or had the star swung a pair of narrow jets on a lazy trajectory across the sky? The view from certain angles, during my approach, had suggested the latter, but it was too late to reconstruct the geometry of the expelled matter with any confidence. As for the layering of the winds, what well-organized structure had once existed was by now washed out by the ravages of time and turbulence. This was an old and homogenized planetary nebula. But it was far from smooth. On close examination, many of the diffuse bright patches proved to consist of aggregations of tiny, dense clumps of gas, presumably condensed out of the supergiant wind. Some of the clumps were so opaque that they harbored molecules and dust grains, which had survived despite the harsh environment. And all the regions of the nebula, clumps as well as the less dense gas that filled the spaces between them, bore the chemical traces of the star's deep interior. There were places where both helium and nitrogen were highly concentrated, compared to their concentrations relative to hydrogen in, say, the Sun. These patches of gas probably came from layers where hydrogen was still being fused into helium, accompanied by the transmutation of oxygen and carbon into nitrogen—one of the subtle games of nuclear physics played in such regions. In other places, the excess carbon was particularly striking. The stellar debris was mixed in everywhere: What the supergiant wind hadn't dredged up in the early stages of the nebula's expansion had been injected forcibly, later on, by the impact of the fast winds.

When I got very close to the central star, within a fraction of a light-year, I finally caught its full glare. The surface was blindingly bright, even in visible light, although nearly all the light came out as very harsh ultraviolet rays. I estimated the temperature of the surface to be well over 100,000 degrees, far hotter than any normal star. But this was nothing compared to conditions a hair's-breadth beneath the surface, where the temperature had to rise to perhaps 100 million degrees to support the nuclear reactions that were still going on in a thin shell. The sharp blue-white sphere in front of me was tiny, only about as big as Earth, and although my view to the nuclear furnace was still blocked, I sensed that at least part of my quest was accomplished. I had seen inside a red supergiant, all the way to the core.

There could not have been much nuclear fuel left to burn. At this close range I finally encountered a fast wind—some thousands of kilometers per second—rushing away from the surface. It carried little mass and must have been a shadow of the gusts that had once swept through the inner regions of the nebula. Such a wind had to be driven by the nuclear reactions just below the surface, where helium was still being fused into carbon and perhaps a small amount of carbon was being transformed into neon and oxygen. But for a star of this mass, most of the carbon would never become hot enough to burn.

I contemplated the future of the Dumbbell's central star. It had just about reached its last gasp of nuclear burning. Soon the veneer-thin nuclear furnace would run out of fuel entirely and begin to cool down. What would then prevent the remaining core, an inert ball of carbon, from collapsing? To ask that was to beg the question. This tiny core, which weighed nearly as much as the Sun, was already too cool to support itself against its own gravity, even as it played out its nuclear endgame. Something else was preventing it from shrinking, and I knew what it was.

I remembered puzzling over the equilibrium that allowed neutron stars to resist their enormous gravitational fields. I had pondered this while the harsh metallic glare of the Crab II pulsar beat against *Rocinante's* skin. There, it had been the atomic nuclei, crushed down to pure neutrons, that had resisted further compression by virtue of their proximity to one another. It was the pressure arising from the bizarre quantum mechanical effect known as degeneracy. Here the same principles were at work, but it was the degeneracy of the electrons that resisted collapse. Gravity in the core of the Dumbbell was squeezing the electrons so close together that they had no choice but to speed up in a chaotic dance. This motion, though not at all related to temperature, was adequate to prevent the collapse of this star, for now and forever into the future. It would never grow much smaller than it was now, and although it would gradually cool and fade, billions of years would pass before it disappeared completely. For the foreseeable future it would remain white hot—the kind of body astronomers call a white dwarf.

Whatever carbon and other trace elements were now locked up in the core of the Dumbbell Nebula's star would remain there permanently. This material had been taken out of circulation, and it hardly mattered into what chemical elements it had been transmuted. But on balance, this star had returned more than it had kept. Nearly the whole envelope—a quantity of matter larger than the core that remained behind—had been processed into a mixture rich in fresh elements that had been dispersed into interstellar space. The Dumbbell Nebula, already quite spread out, would eventually lose its integrity, its atoms mingling widely with the atoms that had been expelled from other stars—or with gas that had never formed part of a star—elsewhere in the Milky Way Galaxy. This blend of elements could someday find its way into new stars and planets. Some of the stars of the next generation would grow into supergiants, create planetary nebulae, and further enrich the Galaxy's chemistry.

Thus planetary nebulae, and their red supergiant progenitors, proved to be the factories that produce and disseminate most of the fresh carbon and nitrogen in the Universe. Whenever the substance of such a nebula was incorporated into a new star, that star would have a higher concentration of these elements, compared to the primordial element hydrogen, than the stars that had preceded it. If a star's matter had been recycled through multiple generations of stars in the past, then it would have that much higher a concentration of these elements.

Does the thought that stars are not made of the same raw materials, one generation to the next, give you pause? To me this realization was no more or less astonishing than the understanding, brought home so strikingly in Orion, that there are generations of stars—that the census of stars and planets is not fixed. All kinds of new structures are always being formed, and some last only a short time, by cosmic standards, before dissolving and freeing up their matter to form something else. If they cycle through life and death, then why not grant them evolution, too?

My journey had evidently shifted, from one of discovering how things are to one of perceiving how they change and evolve. As time was passing for me, and much more time was passing for those I had left behind on Earth, so was it passing for the Milky Way. There was no destroying these newly formed elements, no going back.

Investigating Betelgeuse and the Dumbbell Nebula had not completely satisfied my curiosity about the relationships between stars and their environments. It had not shown me where the other heavy elements—oxygen, magnesium, silicon, sulfur, iron—came from, although the theories told me that stars like Betelgeuse (the massive stars) would provide the key if I waited long enough. I would soon confirm this for myself, under harrowing circumstances. Nor did it demonstrate the entire evolutionary cycle, of star birth, death, and rebirth, all in one place. The matter of the Dumbbell Nebula would merge silently with the rest of the Milky Way, that much was clear. But who knew in which quarter of the Galaxy its atoms of carbon or nitrogen would next join a star's envelope or a planet's atmosphere?

My mental picture of the Galaxy, which had started out so simple, was now becoming criss-crossed by a cat's cradle of interconnections. The matter liberated in places like the Dumbbell, over here, affected the formation of stars and planets in places like Orion, 2000 light-years away. The same principles, driven by gravity, motion, and the concept of equilibrium, now punctuated by evolution and the incessant emergence and dissolution of structure, kept reappearing in every place I visited. Yet somehow it was difficult to put it all together. The Milky Way Galaxy was beginning to feel too big to comprehend. I sought a more self-contained setting that felt more like a neighborhood. The Magellanic Clouds seemed just the place.

25

Leaving Home

My motives for visiting the Magellanic Clouds were anything but simple. First, I sought relief from the vastness of the Milky Way's disk, the complexity of which was beginning to overwhelm me. In the Magellanic Clouds I hoped to be able to sort out the complicated interrelationships that had impressed themselves on me of late. In my travels so far I had seen hints—more than hints—of cycles of stellar birth and death, great currents of mass and energy shaping structures that were too large to grasp mentally, let alone in one's real field of vision, and subtle evolutionary trends whose significance for the Galaxy's development were unclear. In short, I needed some R&R, the opportunity to reflect on these ideas without being forced to face any new ones. The Magellanic Clouds had some of everything—vast star-forming regions bigger than Orion, planetary nebulae, globular clusters, even some incipient spiral arms—all tied up in a couple of compact packages that could be comprehended as a whole, or so it seemed. In adopting this view, however, I turned out to be hopelessly naïve. The Magellanic Clouds were no more self-contained or independent of the entire Milky Way system than the Orion Nebula was independent of Gould's Belt.

At least I could take in both Clouds in a single visual panorama, as I had done on my first foray deep into Earth's

Southern Hemisphere. I remember thinking what a particularly European bit of chauvinism it was that the Clouds had been named after Magellan. These patches of soft fluorescence are every bit as striking as the band of the Milky Way, perhaps more so because of their isolation in the sky. They figured prominently in the celestial mythologies of the indigenous peoples of the south, long before European traders plied those seas. And because there is no southern counterpart to Polaris, no southern pole star, they had served as guideposts to sailors—from Europe and elsewhere—long before the time of the Portuguese circumnavigator. They seemed as isolated in three dimensions as they did in two, and thus they provided me with just as clear a guiding beacon.

Next, there was a more technical rationalization for the visit. I knew that the stars of both Magellanic Clouds had been surveyed by generations of astronomers before I left Earth and that they exhibited certain interesting peculiarities. The most important was that their concentrations of heavy elements—iron and oxygen, in particular—seemed low, as though the Clouds had indulged less vigorously in multiple generations of star formation and recycling than had the disk of the Milky Way. Gazing out of the disk and toward the Clouds I found this puzzling, because the Large Magellanic Cloud, especially, seemed to be rife with massive young stars. One nebula in particular—labeled on my charts "The Tarantula," although it didn't look any more like a tarantula than the Crab Nebula looked like a crab (What was it about arthropods that fascinated early astronomers?)—put Orion to shame. If the Large Cloud had ever known anything like this level of star formation in the past, it should have been richly supplied with the elements cooked in the massive stars' furnaces. The Small Magellanic Cloud seemed calmer, its star clusters older and more sedate. Reassuringly, its abundances of the heavier elements were even lower than those measured in the Large Cloud. Could I trace the origins of the peculiarities by examining the Clouds close-up?

Finally there was a thrill I hadn't anticipated, one that sneaked up on me as I turned *Rocinante* in the direction of the

Clouds. For the first time *I was heading out of the Milky Way.* All of my paths so far had been confined to the narrow plane of the Milky Way's disk. Occasionally I had bobbed up and down across the molecular cloud deck, for a better view. But I had never strayed more than a few hundred light-years from the plane that marked the Galaxy's equator. To reach the Magellanic Clouds, I would have to take off at about a 45-degree angle with respect to the disk and keep going. The distance would be a long jump—160,000 light-years each way, more than 6 times the distance from Earth to the Galaxy's center—although it would add less than 24 years to my travel time, and still less to my age, thanks to the Shangri-La factor and the benefits of hibernation.

Most astronomers regarded the Magellanic Clouds as separate galaxies, distinct from the Milky Way. But this was only partially true. The Clouds were certainly well separated from the Milky Way's *disk* and contained a microcosm of nearly everything the disk contained—stars, gas, the works. Though smaller than the Milky Way (each contained only a few percent of our Galaxy's mass, if that), they could hold their own as respectable galaxies. Yet the Magellanic Clouds were captives of the Milky Way. They lay entirely within the Milky Way's halo, that great domain of faint stars and hot gas in which the disk was also deeply embedded. Had the Clouds once been truly independent star systems that had strayed too close and been captured? Or had they always been destined to merge with the Milky Way? Were they merely bits of the disk that had been unaccountably delayed in joining the Galaxy? In any case, their fates were now sealed. Their presents and futures were being irrevocably shaped by the hostile environment they encountered as they flew through the outer reaches of our Galaxy and by their interactions with one another. In the end there would be no escape.

Traveling toward the Magellanic Clouds entails a different kind of drama than traveling along the plane of the disk. The molecular thunderheads, atomic hydrogen cloud-decks, and dense network of warm wisps are all left behind within the first 2000 or 3000 light-years. The chimneys of hot gas, which form a warren of tunnels and bubbles through the denser clouds, expand and coalesce, and one soon emerges into open territory. The organized circular motion of the disk is left behind; here the stars move chaotically, every star for itself. The gas that fills the space between the stars is so hot that the Galaxy's gravity seems to have little effect on it and so transparent that it is barely detectable except for a pale X-ray glow.

As I accelerated away, the structure of the disk below unfolded at a tremendous rate. I saw the row of atomic hydrogen clouds forming a broad curve, pushed up against the disturbance of the Orion spiral arm like the banks of clouds thrown up against a coastline by on-shore winds. Behind them the dark and billowing molecular clouds reared up where the hydrogen clouds coagulated. The giant molecular cloud complexes were outlined in silhouette by bluish light scattered around their jet-black edges, the light emanating from hidden regions of star formation, unseen Orions. Its bluish tint was imprinted by the smoky haze of interstellar dust. Shafts of intense pink light—the light of ionized hydrogen—momentarily gleamed through holes in the clouds. As my position changed, these beams came and went so quickly, and through such a complex maze of chinks and gaps, that I found it hopeless to try to identify any of the famous nebulae. For a moment I caught a particularly intense field of pink and green and spotted the flash of three or four white-hot stars in a compact package. Was it the Trapezium cluster? It was gone from view before I could check its position.

The patterns were easier to follow when I looked away from the opaque concentrations of molecular gas. Supergiants were easy to pick out: the rubies, Betelgeuse and Antares, and the diamonds, Deneb and Rigel. I spotted some ordinary red giants—Arcturus, Capella, Aldebaran—and the sharp, colorful disks of a handful of planetary nebulae. Other expanding shells of gas, more ragged than planetaries, marked the sites of violent stellar explosions. Ordinary stars, including our undistinguished Sun, formed a richly textured and multicolored backdrop, filling half the sky from my vantage point.

The immense pinwheel pattern of the Galaxy roared into view. The disk was amazingly flat and thin here and bore a thick and even peppering of stars, but the superimposed spiral pattern of dark cloud banks and bright rims of star formation gave it a deeply incised look. Toward the outer Galaxy, I could see the spiral arm of Perseus and a few regions of star formation beyond. Somewhere out there the disk petered out in a mush of indistinct hydrogen clouds, thickening and warping out of the plane for reasons not entirely clear. Toward the Galaxy's center, I made out the Sagittarius arm and beyond it the tightly wrapped bands of atomic and molecular gas pushed aside by the tumbling stellar motions I had encountered on the earliest leg of my journey. The Milky Way's bulge reared up 20,000 light-years in the distance, encircling the hidden Galactic center and fading gradually into the halo. Studding the bulge were dozens of globular clusters, those marvelously condensed spheres—hundreds of thousands of stars apiece, and only a few light-years across—that were very likely left over from the Galaxy's primordial distillation.

I appreciated the contrast this glorious view made with the relative featurelessness of the halo. There were many stars in all directions, but except for the few globular clusters that orbited the Galaxy this far out, they did not congregate in clusters. Nor were they particularly bright. These were all old stars, formed so long ago that even the ones slightly heavier than the Sun had run out of fuel. They shared (and exhibited to a much greater degree) the Magellanic Clouds' peculiarity that heavy elements were unusually scarce, a fact that made more sense here than in the Clouds because these regions clearly had not benefited from stellar recycling any time within the past few billion years. Bright beacons out here were sparse but welcome sights. There were no supergiants, because these would have burned out long ago, but I noted some yellow-white stars, which must have been fusing helium in their centers, and a few red giants.

Occasionally, a small, isolated cloud no more than a few light-years across would appear out of nowhere and startle me. Once or twice I passed straight through such a cloud. For the most part, these clouds were of mild temperature and consisted of individual hydrogen atoms with the usual admixture of dust, but a few of them contained dense knots rich in molecules. Were they shards of matter falling into the Galaxy for the first time, or had they been ejected from some explosion in the disk? I could not decide, but I quickly realized that these would not be the last of the clouds I would have to contend with. Like a pilot who climbs through a cloud-deck only to find an inaccessibly high layer of cirrus far above, I perceived that I was heading toward a more ubiquitous layer of atomic hydrogen at about the distance of the Magellanic Clouds.

26

Outposts in the Halo

From my location 100,000 light-years out in the halo, I could see the Clouds' three-dimensional structure. They were highly elongated, and (in a curious echo of my trip through the bulge of the Milky Way, en route to its center) I noticed that many of their stars shared the kinds of stretched-out and figure-eight orbits that gave these galaxies the appearance of tumbling through space. They were not tumbling in unison; however, they were flying in formation as their orbital trajectories carried them around the Milky Way. The Small Magellanic Cloud, farther distant by maybe 20,000 or 30,000 light-years, was pointed away from the Milky Way's disk, whereas the Large Cloud presented residents of Earth with a view much closer to side-on. The bright regions of recent star formation, especially in the Large Cloud, were asymmetrically placed with respect to the main bodies of stars, making the clouds look lopsided.

The vast cloud of cool hydrogen, which I had noticed as I traversed the halo, not only enveloped both Clouds but also spread out in a seemingly interminable stream to either side of them. I had heard about the Magellanic Stream, because it was visible to radio astronomers on Earth. Its significance had puzzled my colleagues for years. Did it consist of matter stripped from the Clouds and left behind as they plowed through the Galaxy's halo? Or had the gas been teased from both Clouds by their gravitational attractions for one another?

The trajectories of the Clouds seemed clear enough. Their orbits about the Milky Way had brought them in from a much greater distance—maybe 3 times farther out than they were now—and they were skimming along at roughly their minimal separation from the Galaxy's center before heading back toward the outer halo. Maybe they had made half a dozen passes like this, each one closer than the last as the resistance of the halo's stars and gas took its toll. It made sense that they should leave behind a smear of liberated gas as their orbits eroded and they spiraled slowly toward their destiny of complete absorption by the Milky Way.

But the theory that the stream traced out the Clouds' orbital path had a serious flaw: The "trail" extended in front of the Clouds as well as behind. Friction of the halo against the Clouds' motion would certainly not draw gas out ahead of the Clouds, but gravity might. Whose gravity, though? The Milky Way's gravitational field was too gentle to extract so much gas from the Clouds. As galaxies in their own right, the Magellanic Clouds had a strong propensity to hold on to what was theirs. But close encounters between the Large and Small Clouds were assured and must have occurred not too long in the past. These could be much more disturbing to the Clouds' interiors. Could the stream, then, be the result of devastating tides raised by the Clouds on one another, when they had met fatefully hundreds of millions of years earlier?

This theory had its problems, too, although it seemed more promising. Try as they might, astronomers had never been able to detect the stars that should have been pulled out of the Clouds along with the gas, and I couldn't, either. Gravity affects stars and gas equally, and it would be hard to imagine a gravitational interaction that would extrude so much gas from either galaxy without also liberating quite a few stars. My ability to fly around the Magellanic Stream was little help to me in understanding its origins. Fortunately, much less mystery surrounded the origin of the matter that bridged the gap between the Clouds. This gas contained a reassuring complement of stars and seemed certainly to be remnant of one or more close encounters. One of these encounters must have been very recent, because a curved streamer of stars, the very signature of death by gravitational tide, bulged out of the Small Cloud and intruded halfway across the bridge, still intact. The Large Cloud seemed to be pulling the Small Cloud to pieces.

The peculiar patterns of star formation in the Clouds also seemed to reflect the injuries that these galaxies must have inflicted on each other in the past. Why else would star formation have been dead for nearly the entire history of the Clouds and then suddenly burst into life only a couple of billion years ago? It seemed a likely explanation for the paradoxical conjunction of rapid star formation with a scarcity of heavy elements. The close encounters must have triggered star formation—especially in the Large Cloud, which had held on to more of its molecular gas. I recalled the evidence I had seen for sequential bursts of star formation across the Orion region, where the birth of stars in one location had triggered a subsequent burst of star formation nearby. The mechanism in Orion had been the shock waves, rolling out of the region of massive star formation in one newly formed cluster and overrunning a neighboring molecular cloud, squeezing it and pushing it over the edge of collapse. But a collision between the Clouds: That could trigger an Orion on a massive scale. Here the shocks would carry not merely the momentum of the exploding and wind-producing massive stars but the full impetus of two colliding galaxies.

Once again, I was becoming overwhelmed by details and shaken by the realization of how interconnected these cosmic structures are. My search for a simple, "closed" system (to get away from the complications of the Milky Way) had brought me to a pair of galaxies that were so open to environmental influences that they were literally at the mercy of their surroundings and of each other. Recycling of matter through stars and the inexorable creation of heavy elements was one form of evolution.

Here was another, even more dramatic form of evolution—a form that would continuously shift the shapes and contents of entire galaxies and eventually dissolve them into one another. *Even a galaxy could not be regarded as a closed system.* The effects of all this external buffeting trickled down to the internal structures of the Magellanic Clouds themselves. Had the Small Cloud failed to produce even the paltry complement of oxygen and iron found in the Large Cloud? Perhaps this was because its element-enriched gas had been stripped away before the galaxy had had a chance to recycle it. The Large Cloud had apparently been more effective at holding on to its gas, and one could foresee the day when it might absorb the Small Cloud, before making its final death spiral to join the bulge of the Milky Way. But even the Large Magellanic Cloud had been quiet for many billions of years, until a collision with the Small Cloud had released its pent-up star-forming potential and caused it to burst into glory.

Dramatic evidence of these influences lay right in front of me, less than 10,000 light-years away. I could not take my eyes from it. Off to one side of the galaxy and amid a scattering of lesser sites of star formation, the Tarantula Nebula shone with the brilliance of several thousand Orions. Orion had its Trapezium of four massive hot stars; the Tarantula had thousands. Weary of generalities, I set my course for the Large Magellanic Cloud, eager to immerse myself in what had to be one of the most glorious environments this side of Andromeda. What I got instead was another lesson in the evolution of stars.

27

The Explosion

Somewhere in the observable universe—by this I mean the Universe within a distance of about 10 billion light-years, within which light can have reached us since the Big Bang—a star explodes every second. But even the observable portion of the Universe is a very big place, and stellar explosions are consequently considered rare events. Thus, when I came to the Large Magellanic Cloud, the last thing on my mind was a life-and-death struggle to outrun the blast from a supernova.

Even before I approached the Cloud, my attention had been drawn to a very bright star that seemed to be behaving strangely. I had noticed it not long after leaving the Milky Way's disk, because it was so bright and because it was situated in the middle of a little dark hollow at the edge of the Tarantula Nebula. As I watched, the star changed color from orange-red to a hot blue-white. At this point, my velocity was within 1 part in 10 billion of the speed of light, and events outside my craft appeared speeded up by nearly a factor of 100,000—that is, by my Shangri-La factor. But I was able to calculate that the transition had taken only a few thousand years of Earth time (incredibly fast for a stellar transformation). I should have sensed danger, but I was intrigued. I set a course for this star, thinking that it would be an interesting base for explorations of the Tarantula.

Pulling up nearby, I confirmed the classification I had hazarded en route. This star, in its present incarnation, was what is known as a "blue supergiant." The blue coloration reflected the temperature of its surface layer, which was not even 4 times that of the Sun. I estimated it to be about 20,000 degrees, maybe a bit more. I guessed that its mass was about 18 or 20 times that of the Sun, not very different from Betelgeuse. I was well aware that the brightness of a stellar surface went up very steeply with temperature (as did the brightness of any other opaque surface); this fact, and the star's size (27 times the diameter of the Sun, but only 1/50 the size of Betelgeuse) accounted for its huge luminosity: 100,000 times the Sun's output. Being so far from home and slightly anxious about having ventured outside the Galactic disk for the first time, I thought it would be pleasant to bask in starlight at just about the level that Earth does in its orbit around the Sun. To compensate for the higher luminosity, this meant staying much farther away. (*Rocinante* was sufficiently well shielded that I could have parked closer in.) Little did I know that this nostalgic whim would save my life. After circling a bit to get the best view of the Tarantula Nebula, I parked my craft at about 300 times Earth's distance from the Sun and settled in for a well-deserved nap.

The "incident" began while I was still asleep. I woke up with a start, thinking that someone had shone a laser beam in my eye. I almost convinced myself that it was stinging but decided that this was my imagination, perhaps the tail end of an unpleasant dream. After a few moments there seemed to be no ill effects, and I went back to sleep.

No one had shone a laser beam in my eye, of course; this is not a science fiction story. But it wasn't my imagination, either. What had actually happened was much scarier than a crew member gone mad, loose aboard *Rocinante* in deep space, wielding a deadly laser. What had happened was that an eruption of energetic particles—electrons, positrons, protons, gamma rays, you name it—had exploded inside my eye. The reason? An energetic neutrino or two had slipped in sideways, through the supposedly impenetrable wall of my craft, through my skull, diagonally across my frontal lobe, and across my eye socket. After all that, when it was about to get away scot-free, it had a head-on collision with a carbon atom in my ocular jelly and stopped dead in its tracks, dumping its energy *in situ*. And this didn't happen once or twice, it happened about 3 million times in the space of 10 seconds. Only I didn't know that at the time.

Two hours later, all hell broke loose. Without additional warning, the surface of the star exploded. Fortunately, *Rocinante's* shielding protected me from the first flash of gamma and X-rays. Then an initial blast of ejected matter rushed past at close to the speed of light. Lucky again: This barrage, which consisted of just the thin outermost layers of the star, it packed little punch and washed over *Rocinante* without doing much damage. But a much more ominous, dense, tidal wave of debris was rushing toward me at 15,000 kilometers per second, 1/20 the speed of light. I had to get out of there immediately.

The problem was that it took time to accelerate to a speed high enough to outrun the blast. Despite the sophistication of my propulsion system, I was still subject to all the laws of physics, including the one that said that accelerating at a rate much higher than Earth's gravitational acceleration would feel exactly like being crushed under an enormous weight. The old test pilots and astronauts had shown that humans could survive 8 or 10*g*'s for short periods of time, but my craft would not even approach those limits. Because I had not anticipated the need for rapid escape, *Rocinante*—which normally accelerated at 1*g*, the acceleration of a falling body on Earth—was limited to a maximum of only twice that . . . if it worked to specifications. If everything worked, it would take 9 days to reach the speed at which the blast was approaching. If I stood still, the blast would overrun my craft in a little over a month. For an instant, I considered playing "chicken"—banking on the 26 days' leeway I had under the assumption that my technology was flawless. Even if my craft managed only 1*g*, I would still have 17 days at my disposal for sightseeing before I had to flee. Then I thought better of it, turned tail, opened up the throttle and hoped for the best.

Meanwhile, the blast was expanding in my field of view at an alarming rate. Before the explosion, the star had appeared only 1/10 as big as the Sun did on Earth's sky. (This star was much larger than the Sun, but I was parked at nearly 10 times the orbital distance of Pluto.) Only 4 hours after the explosion, the expanding blast appeared as big as the Sun. After 2 days, it was more than 10 times bigger. When I finally reached the speed of the exploding material (Everything worked!) I had traveled only a distance equal to 38 times the distance between Earth and the Sun—a shade more than 10 percent of my original distance from the exploding star. I could have turned off the acceleration at this point and coasted along with the explosion, but the result would have been uncomfortable. The explosion now took up as much space on my sky as the Big Dipper did on Earth's, and it was emitting as much light as a *billion* Suns. Under this cosmic broiler, *Rocinante's* skin was heating up to 3000 degrees, which meant that it was beginning to vaporize at a furious rate. This was still too close for comfort. I continued to pull away, cutting my acceleration back to 1*g* only after the temperature began to subside.

Safe for the time being, I reviewed my thoughts about what had happened inside this star. The interior of a massive star—one that is at least 8 or 9 times as massive as the Sun—never settles down to form an inert core that stays inert for very long. The degeneracy of its electrons is never quite enough to support it against gravity. Every time it uses up one type of nuclear fuel, it continues to shrink, and get hotter, until the next level of fuel ignites.

I guessed that Betelgeuse had already used up most of the helium in its core by the time of my visit. Even during its helium-burning phase, it would have burned hotter than a less massive star like the one that had produced the Dumbbell Nebula. Three heliums stuck together make a carbon atom, but at such high temperatures the violence of the reactions readily could have added a fourth helium, making oxygen. Thus massive stars like

Betelgeuse—and the star that had just exploded—part company with their lighter counterparts as they create a mixture of oxygen and carbon in their cores.

Still there is no respite. The core is shrinking and getting hotter. Not only are the atomic nuclei slammed together with greater force, but they are also slammed together more frequently, and the nuclear reactions accelerate. Each stage of thermonuclear fusion runs faster than the one before. If it took 1 million years to fuse the helium into carbon and oxygen, then it would take less than 100,000 years for all the remaining carbon to be destroyed and the core to be converted to a mixture of oxygen, neon, and magnesium.

It was sometime during this frenzied episode of nuclear incineration that the star in the Magellanic Cloud changed color from red to blue. The exact reasons for the change remain obscure, except that, as in Betelgeuse, there was a serious and growing disconnection between the shrinking core and its distended envelope. Just as slight hiccups could cause large portions of a supergiant's envelope to fly off into space, so could they cause the envelope suddenly to deflate, and this is apparently what had happened to the doomed star in the Large Cloud.

Oxygen, neon, and magnesium fuse into silicon and sulfur, and thence into iron. The last stage, which takes only a few days, must have occurred as I eased my craft into its orbit about the star, oblivious to the drama inside. The creation of the iron core is a watershed event for the star. Iron is not nuclear fuel. No energy can be released by combining the nuclei of iron atoms with other atomic nuclei. Yet the core, too heavy to support itself, continues to shrink and get hotter. Now gravity steps into the limelight, relegating nuclear power to second place. No longer is it the supporting player that inexorably moves the core from one stage of nuclear burning to the next. From now on, gravity calls the shots, and the nuclear reactions follow.

As gravity overwhelms all forms of resistance, the core implodes, at first gradually, then at breakneck speed, contracting the last few hundred kilometers in less than a second. The nuclei of iron atoms break apart under the extreme pressures and temperatures. This sphere of iron the size of Earth, formed painstakingly through the elaborate successive stages of nuclear fusion, becomes an ultradense globe of indeterminate composition, a soup of nuclear fragments, less than 100 kilometers across. Reversal of the core's nuclear alchemy does not stop even with the disassembly of matter into pure hydrogen. The protons and electrons of this primordial element are squeezed together with such vehemence that they merge to form neutrons, the building blocks of the neutron star that is soon to form.

It was at this point that I received my dose of neutrinos. The particles known as neutrinos—Enrico Fermi's "little neutral ones"—are famous for being nearly impossible to detect. They hardly interact with anything, and that's why they were able to pass through the otherwise highly protective shielding of my craft. If I had wanted to build a craft that would protect me from the bath of neutrinos, its skin would have had to be a light-year thick to catch all of them! Yet so many neutrinos flowed out of the imploding core (a billion trillion trillion trillion trillion, one for every proton–electron duo that merged to form a neutron) that 3 million of them managed to come to grief inside my eyeball, snapping me out of sleep. The same thing happened in every other cubic centimeter of my body, but I was lucky. This was not a lethal dose; it wasn't even harmful. If I had been as close to the star as Earth is to the Sun, instead of 300 times farther away, I would have been done for.

The neutrinos had a much more devastating effect on the star. The collapsing star's interior presented such a formidable barrier that even the ultra-elusive neutrinos had a hard time escaping. Most of them eventually made it, but at the cost of pushing so vigorously against the star's interior that the pressure built up to dangerous levels. The core continued to collapse, releasing neutrinos at an accelerating pace as it shrank the final tens of kilometers down to the city-sized dimensions of a neutron star. But the surrounding shells of partially fused matter and the extensive envelope of hydrogen and helium could not resist the accumulated pressure of all those neutrinos struggling to get out. These portions of the doomed star were blasted away as the dammed up neutrinos finally broke free. It took two hours for the explosion to organize itself and reach the star's surface, and only when it burst through the surface was I alerted to flee.

28

Aftermath

In the chaotic supernova blast, the shells of material that hadn't collapsed with the core were stirred vigorously into the exploding envelope. What rushed out at me was composed largely of hydrogen and helium, because most of the star's envelope had never fused, but it was streaked with intrusions of all the partial and final products of nuclear fusion that had layered the doomed star's interior with an onion-like structure. Carbon, which had underlain the hydrogen-burning shell, emerged, along with nitrogen and fingers of gas rich in oxygen, neon, and magnesium. Silicon, sulfur, calcium, and a host of other elements were shot through the exploding envelope from below. Most of the iron was gone, having been sucked into the neutron star and destroyed. But hidden below the surface of the expanding blast and traveling along with it were globules consisting of more exotic species, dominated by a radioactive form of nickel. These elements had been forged in the heat of the explosion, and they would have a curiously important effect on the appearance of the blast during the months that followed.

The blast continued to brighten for about 3 months. By this time I was 1700 times farther from the center of the explosion than Earth is from the Sun, more than 5 times farther than I had been when the explosion went off. Debris still hurtled outward, unimpeded, and would reach this location only 3 months behind me. Should I keep running or allow the debris to overtake my craft? I calculated that after expanding this far, the shrapnel would have thinned out considerably and would present little hazard. It would rock my craft with less than one Earth's atmosphere of pressure—I had withstood more while traveling to the center of the Milky Way at high Shangri-La factor. True, the debris was full of radioactive material, but *Rocinante's* skin would easily protect me from its dangerous emissions. I parked my craft (taking the requisite month and a half to decelerate) and sat back to watch the show.

The brightening was not due to any extra source of power inside the sphere of debris. When the explosion had started, the interior of the blast had been unimaginably hot and full of radiant energy. But that energy could not escape immediately. It was trapped by the immense amount of material it would have to penetrate. The intense luminosity I was seeing now was the residual heat of the explosion finally making its escape, though in a considerably diluted form.

Once this heat began to leak out in earnest, the appearance of the explosion changed dramatically. Its opaque "surface" thinned out, and I could see farther and farther into its interior. The featureless disk that the explosion had presented on my sky, which had cooled down to a temperature and color similar to the surface of the Sun, dropped behind the onrushing front of debris. The latter developed a tortured, mottled texture, with dark spaces opening up between writhing, entangled filaments of light. It looked like an exploding ball of snakes. These filaments were not radiating the smooth distribution of colors that had characterized the opaque envelope but, rather, assumed an array of pure hues that I recognized as the signatures of specific atomic disturbances. Different filaments, having different elemental compositions, produced diverse arrays of color, but the gradations were subtle. The dominant color, as in so many other places I had visited, was the pink of hydrogen—not surprisingly, because this was the dominant element of the star's envelope.

Hydrogen's glow not only outlined sharp striations within the debris globe but also provided a pervasive undercoat in the interstices between filaments. A similar yet distinctive shade of red came from oxygen atoms, confined mainly to certain filaments that must have punched their way through from below. In still other locales, I could detect (with the aid of my instruments) an intense infrared glow associated with calcium. The color contrasts were not so stark as they had been in the Dumbbell Nebula. Here there was no hot star at the center to stir up the atoms with a steady injection of ultraviolet rays. And if the neutron star had become a pulsar, like the one at the center of the Crab Nebula, its emissions were nowhere to be found.

Though I detected many other atomic signatures, their signals were weak. The breakneck expansion of the debris had apparently chilled it to the point where it was too cold to extract much emission from helium and many other elements I knew to be present. The original heat of the explosion was now mostly gone, and one might have expected the brightness of the exploding debris to fizzle out entirely, at least until it collided with whatever matter surrounded the site of the explosion. But something intervened to keep the supernova shining. After a sharp peak of luminosity and a brief, even sharper plummet, the collective brightness of the debris began to level off and to decline more gradually. Something inside the debris sphere was now providing extra energy, at a very measured rate.

The culprit was radioactivity. Most of the radioactive nickel that had been created in the moments following the explosion, forged at temperatures of 200 billion degrees or more, had decayed away within the first couple of weeks. Its decay had released some energy, but not enough to compete with the energy of the explosion that had been slowly leaking out at the time. However, the sizable amount of nickel—nearly 7 percent of the Sun's mass—had not simply vanished, nor had it decayed into something benign. It had simply changed into another radioactive species, a form of cobalt. Radioactive cobalt's half-life (the time required for half its atoms to disintegrate into stable iron) is

77 days, so there was still plenty of it left after 3 months. Even 6 months after the explosion, the decay of radioactive cobalt, and the gamma rays it emitted, powered the glow from this huge and growing blast.

Right on schedule, the brunt of the debris overran my craft. The jolt was rougher than I had expected—there were some compacted blobs that hit *Rocinante* with considerable force—but there was little physical risk and no damage. I was thankful that the debris had thinned out to such a degree that I could see most of the way across the expanding sphere. It would have been unnerving to be caught in the opaque fireball with nothing but a uniform glow on all sides, even if the temperature had cooled down enough to be survivable. Still, the glowing filaments rushing toward and around *Rocinante* at 15,000 kilometers per second created an eerie sensation, very different from the sensation I had experienced (as a consequence of *my* motion, in that case) as I traversed the Orion Nebula, back to front.

I could now appreciate the role that radioactivity played in lighting up the ejected material. The nickel had not been spread uniformly through the debris but had been concentrated into globules within the inner 20 percent, or so, of the expanding sphere. When it decayed into cobalt shortly after the explosion (its half-life is only 6 days) the energy it released was trapped and could not even spread smoothly through the debris cloud, much less escape from it. As a result, the nickel-rich globules became hot compared to their surroundings, and "popped" like popcorn, pushing aside the material that surrounded them. What had started out as compact globules, rich in radioactive nickel, became "holes" full of radioactive cobalt, and the inner part of the debris cloud assumed the texture of Swiss cheese. Now that the cobalt was decaying, these holes were glowing in gamma rays.

The spectrum of radiation inside this explosion was growing weirder by the day. There were the gamma rays emitted by decaying cobalt, of course, a kind of radiation usually associated with very hot gas. These energetic photons were hitting electrons and bumping them up to high speeds, whereupon the electrons produced X-rays. At the same time, the gamma rays, fast-moving electrons, and X-rays were knocking into the atoms of the debris, creating the atomic disturbances that led to the assortments of colors I have already noted. Yet amid all this activity the gas itself remained rather cool, hovering at only a few thousand degrees. When I was first submerged by the blast, many of the atoms were being driven to produce colors in the visible part of the spectrum. But with time, the level of disturbance declined and the dominant "colors" drifted into the infrared. Some denser pockets were getting down to lower temperatures, and I began to notice indications that molecules were forming. Dust, mostly graphite, was also beginning to condense out of the cooling envelope, just as it had done in the winds from the red supergiants. Eventually, my pristine view across the sphere of debris degraded as I found myself surrounded by clouds of soot.

All this time, the first flash of light from the supernova was racing out into space. Traveling at the speed of light (of course), the flash was outrunning the physical debris by 20 to 1, but so far there had not been much in its path to catch the rays. That changed about a year into the explosion, when the light reached the inner edge of the wind that had been expelled from the star when it was a red supergiant. The supernova now began to light up its surroundings.

I had anticipated that there would be a year's delay or so between the flash of the explosion and its echo against the matter that surrounded the exploded star. From my own observations as I approached the Large Magellanic Cloud, I knew that the star had changed from red to blue some 3000 or 4000 thousand years earlier. Blue supergiant stars, of course, produce a wind that is, if anything, more powerful than the wind from a red supergiant. But such winds impress via their speed, not via the amount of matter they inject into the region immediately surrounding the star. Thus the light from the supernova could pass straight through the relic of the blue supergiant wind without producing much effect. The red star's wind, however, would have been slow and dense, and when the intense light of the supernova finally reached the remnant of that wind, it would surely light up like a neon sign.

In 3000 or 4000 years, the matter expelled by the red supergiant, with a typical speed of 20 kilometers per second, would have traveled about a quarter of a light-year. Once the star turned blue, though, its wind must have quickly overtaken the slower wind from behind and hurried it along, so I guessed that the interface would be located somewhat farther out, maybe close to a light-year. The echo would then take nearly a year to get back to my position, so I expected that I would see the sky light up just before the second anniversary of the explosion.

Witnessing the first assault of the supernova on its surroundings would provide a kind of closure to this leg of my journey. The initial flash of the supernova's light had been intense with ultraviolet rays, and if the dense gas expelled by the red supergiant surrounded me on all sides, then the whole sky should light up in the exquisite green of twice-ionized oxygen. The contrast with the deep reds of the expanding debris promised to be spectacular, and the symbolism was not lost on me either. A slow wind swept up by a fast wind, the flash of the explosion replacing the steadier light of a newly minted white dwarf . . . the analogy to being inside a planetary nebula was almost too perfect. And like a planetary nebula, this supernova was participating in the grand evolutionary cycle of the Universe. The flash would signify the connections between the death of this star and the lives of stars yet unborn. Instead of being responsible for synthesizing carbon and nitrogen, this star's portfolio included oxygen, neon, magnesium, and silicon, all mixed up with a goodly amount of hydrogen and helium ready for reuse. Only the iron core—the matter that was now to spend eternity as a neutron star—had vanished from circulation, memorializing the exploded star's existence just as white dwarfs constituted lasting monuments to stars of lower mass.

Just shy of 1 ½ years after the explosion, the flash appeared. It did not look as I had expected it to. Instead of lighting up all over the sky, as it would have done had the red wind formed a spherical container around the blue wind's bubble, the green light appeared as a narrow band, maybe half the width of the Milky Way as viewed from Earth, which lengthened until it circled the entire sky. There was no spherical cavity, that much is clear. Apparently, the nearest parts of the red supergiant's wind had been pinched into a ring, probably the waist of an hourglass shape that fanned out . . . who knew how far? Once again the connections struck me: the Dumbbell Nebula, with its broad waist, about to fade into the interstellar background; the Ring Nebula; the Saturn Nebula, with its mysterious "ears;" and so forth. I thought all the way back to what I had seen in Orion: the star-forming regions with their dual funnels pouring out light and their ubiquitous jets. The same shapes kept appearing and were repeated here, even in the chaos of a star's violent self-destruction. I knew that at this moment—I mean, of course, when the light finally reaches Earth 160,000 years hence—the astronomers of those times would appreciate the deep connections among these phenomena.

29

Afterthoughts

The explosion at the edge of the Tarantula Nebula would scar the Large Magellanic Cloud for ages to come. Within 100 years, the shrapnel from the star would have plowed into so much of the surrounding matter that it would have begun to slow down. The surroundings, swept into a thick shell, would absorb what remained of the explosion's energy, carrying on the expansion while the spent debris fell behind. But fingers of the exploded star's matter would continue to poke outward into interstellar space, mixing with and enriching the surrounding gas with the fruits of its nuclear alchemy. Eventually, its own peculiar blend of heavy elements would meld with the dollops of carbon-nitrogen mixture being supplied by planetary nebulae that were as common here as they had been in the Milky Way.

But that would take time. In tens or hundreds of thousands of years, or perhaps a million years, there might still be an ultraviolet or even an X-ray reminder of the star that had exploded as I watched. If no fresh catastrophe overran the neighborhood, shredded filigrees of glowing gas would eventually outline a shell dozens of light-years across, which would gradually slow down and fade until it disappeared into the background. Analysis of the atomic disturbances, X-ray colors as well as ultraviolet and visible, would still bear the imprints of chemicals created here. Just then my eyes caught the blinding glare from the thousands of hot massive stars that filled the center of the Tarantula Nebula. Each one was destined for a similar explosive demise, and each would propel the surrounding gas into an echo of its own expansion. Chances were that one of these explosions would push aside and smear out my supernova's remnant long before it had run its natural course. I surveyed the onrushing debris and glowing ring wistfully, surprised to feel a proprietary interest in an event so far beyond the scale of any human endeavor. And I was amused to find myself consoled by the thought that the untimely disruption of my star's remnant would only speed up the incorporation of its heavy elements into the fabric of the Large Magellanic Cloud.

Whatever the fate of the supernova's gaseous remnant, the star's imploded core would stand fast as a neutron star, perhaps shining for a time as a pulsar. But was it really there? By now its effect on the debris should have become apparent. I searched in vain for something giving the debris cloud extra oomph, adding some power to it over and above the decaying radioactive cobalt. If the neutron star were sending out a searchlight beam, producing flashes of ordinary light like the Crab Nebula's pulsar, those flashes should have been visible directly. But there was no sign of any pulsar. It was unthinkable that the entire core had exploded. Something was surely left behind, but it seemed to be invisible. Was there some reason why the neutron star had not begun to pulse? Had something robbed the core of its angular momentum, damping down its spin as it collapsed? Or had its magnetic field failed to develop? Had something more interesting happened? Could enough matter have fallen back onto the core to drive it over the edge, causing it to collapse to a black hole? I briefly considered investigating for myself but decided against it. I was not inclined to enter the nebula any more deeply. Like the debris, my thoughts were turned outward, and I had little inclination to explore the kind of tortured confines near a neutron star—or black hole—that had proved so anxiety-provoking before.

One thing was clear: I had come nowhere close to escaping the complexity of the Milky Way by traveling to its satellites. The Magellanic Clouds were more complex, if anything, because they were *less* self-contained. These galaxies had evolved differently from the Milky Way's disk and from each other. Was it because of their frequent encounters with one another or because of their incessant plunge through the stars and gas that made up the Milky Way's halo? Were the Clouds poor in oxygen and iron because their stars had never manufactured the quantities of heavy elements that larded the Milky Way's disk, or had massive stars once lavished oxygen, silicon, and iron on the Clouds, only to have these elements stripped away along with the gas they permeated? I tried to project the evolution of the Large Cloud into the future, long after the thousands of hot massive stars that filled the Tarantula Nebula had exploded. Would astronomers then say that the Large Cloud had finally caught up to the Milky Way in its chemical richness? Or would much of this oxygen-enriched gas be pulled out of the Large Magellanic Cloud the next time it got too close to the Small Cloud, before it had a chance to form a new generation of stars? Perhaps the gas would be blown out of the Cloud by the force of those concatenated explosions. I wondered whether the outcome of all this stellar activity might be the enrichment, not of the Magellanic Clouds, but of the halo of the hegemonic Milky Way that would soon engulf them. For how long would there be Magellanic Clouds at all?

Having seen the grandeur of a supernova explosion, and having allowed the debris to sweep over me only a few months after the event—having been *inside* a supernova debris cloud—I found further exploration of the Tarantula Nebula and its vicinity hopelessly anticlimactic. I was impatient to move on. I knew that by leaving now, I would miss a spectacle beside which the illumination of the red supergiant's wind would pale. Twenty years after the explosion, the debris shell itself would finally ram the dense ring that formed the waist of the hourglass, and the ring would sparkle brilliantly in every part of the electromagnetic spectrum.

But a close-up view of this marvel would have to wait for another time and another supernova. There were many pressing questions still to answer. Should I continue to pursue the puzzle of evolution or set it aside in pursuit of other, equally insistent goals? I was bothered that big pieces of the puzzle were missing. For example, this supernova, despite its dramatic nuclear endgame, could not be the major source of cosmic iron. Carbon and nitrogen were ably supplied by the calm outgassing of planetary nebulae; oxygen and many other elements were derived from violent explosions like the one I had just witnessed. But most of the iron had to come from somewhere else. By this time the cobalt had just about finished decaying, and I knew that this supernova would not release much more than the 7 percent of the Sun's mass in iron that had been produced through radioactivity. The much larger quantity of iron that the star had created in its last week or so before exploding had been crushed to neutrons and locked up forever. There had to be yet another route by which matter could be forged into iron in a nuclear furnace and then released for future use.

If I wished to seek out the missing iron, I knew where to look. During my long traversal of the Milky Way's halo, I had noticed, far in the distance, 100 or more flashes at random intervals and seemingly at random locations throughout the halo. This journey, which for me had required only a couple of decades of hibernation, had sampled 160,000 years in the life of the Galaxy. Therefore, these flashes were appearing every 1000 years or so. I had had the presence of mind to record the signals from some of these brief flares, and now I knew much more about their nature. They, too, were explosions of stars, but explosions of a rather different sort from the one that had produced my supernova. They were brighter and faster, and they showed no signs of the pink hydrogen light that dominated the explosion I had just witnessed close at hand. I remembered my colleagues debating the provenance of these events during the years before I left Earth. It seemed that they were explosions of stars that had completely lost their hydrogen envelopes. Because they were seen to occur in the Milky Way's halo as well as in the disk, they had to be old stars. People believed that these supernovae were the explosions of white dwarfs.

How, in the calm reaches of the halo, an ancient, dead star could suddenly blow up—that had been a major mystery. It had been supposed that the explosion occurred when so much matter was dumped onto the surface of a white dwarf that the pressure of its electrons could not support the extra weight. In these remote environs, the fresh supply of matter could come only from a companion star. The dwarf would start to collapse, but this time, nuclear reactions would intervene before gravity got the upper hand. The entire star would explode like a powder keg, incinerating its store of nuclear fuels in a great burst and leaving no cinder behind. This kind of supernova produced the iron in the Universe.

I thought how easy it would be for me to determine, once and for all, how these explosions were triggered. Should I strike out into the Milky Way's halo to pursue this third venue for nuclear alchemy? There was the considerable problem of trying to pinpoint a likely candidate for such an explosion. Positioning myself would be tricky, even given *Rocinante's* maneuverability. But there was a more formidable obstacle: I could not work up the enthusiasm for a further foray through the Milky Way, disk or halo. Even though I had never left the Milky Way's halo, my exploration of the Magellanic Clouds had given me a new perspective on the home Galaxy. I now saw it as though from outside. I dwelt more and more on the relationships of one galaxy to another. Galaxies underwent great cycles of birth and death, all the time evolving into systems more complex, more diverse. But they did not do so in isolation.

I pondered where to go next. Distant galaxies beckoned, but the home Galaxy represented a kind of security. The disk of the Milky Way began to seem more inviting. It filled half the sky with its mesmerizing spiral pattern, great rows of nebulae, and dark lanes of cloud. It looked like it would contain everything I might ever want to explore, and I must have lost my nerve momentarily, because the next thing I knew I was heading straight toward the Milky Way's disk. But I did not stop there. In what seemed like a few moments I plunged through the disk, emerged from the other side, and kept on going.

Part Six

Hierarchy

30

Really Leaving Home

There must be something tremendously liberating about leaving one's own galaxy. Some of my colleagues, seduced by the intellectual lures of other galaxies and the way they fit together into even larger cosmic structures, had hardly ever visited the Milky Way—metaphorically, that is—during their entire professional careers. I, ensconced *physically* in the Large Magellanic Cloud, though still in the outskirts of the Milky Way's halo, had felt an irresistible urge to escape. But to where? My journey through the halo toward the Magellanic Clouds had been spared from desolation by the constant presence of the Clouds ahead and the reassuring blanket of hydrogen that enveloped them. I was also kept company during my trip by the other two steadfast companions of the Milky Way: the spiral galaxy that lay in the direction of Andromeda on Earth's night sky and the face-on pinwheel that ordinarily lay beyond the constellation of Triangulum, the carpenter's triangle. Each of these galaxies was huge, almost a twin of the Milky Way. Andromeda, more than 2 million light-years away (15 times farther than the Magellanic Clouds), was specially honored as the Milky Way's partner in a minuet. In their orbital dance, now swinging wide apart, later nearly brushing haloes, these two galaxies set the pace for the modest aggregation of galaxies known as the Local Group.

In Andromeda, I could see at a glance how the Milky Way must look to a voyager far outside its reach. The Milky Way and Andromeda galaxies were similar in many ways—the disk, the dusty dark lanes, the star-forming nebulae, the central black hole. Just as the Milky Way had two substantial attendants in the Magellanic Clouds, so did Andromeda travel with two robust companions in tow. But such similarities made the curious contrasts between them all the more striking. Andromeda's attendants looked nothing like the Magellanic Clouds. Their shapes—perfectly smooth, slightly flattened balls of stars—pegged them as members of the other great class of galaxies, the "ellipticals." They boasted no great nebulae, no chains of newly formed massive stars, no spine of stars that resembled incipient spiral arms or a tumbling stellar bar. Where the Magellanic Clouds were lumpy and ragged in appearance, Andromeda's companions displayed the kind of symmetry and compactness I had hitherto associated only with the jewel-like globular clusters. Yet the companions of Andromeda were thousands of times heavier than any such cluster.

I was clearly being taught another lesson about the diversity of cosmic structure. How was I to know the extent to which this lesson would prepare me for the next stage of my journey? After all, my destination had not even been chosen yet. In the excitement over my explorations of the Milky Way's disk, I had taken for granted the special characteristics that set the Milky Way apart. I was well aware that not all galaxies sported pinwheels, but I had not really thought much about what a galaxy would look like if it had no disk.

Remove the disk from a spiral and you would have a galaxy that was all halo and bulge. This game of "what if" suddenly assumed importance, because at a certain level of description, elliptical galaxies resemble the disembodied haloes or bulges of spirals. Like the halo, they consist mainly of older stars, a fact given away by their reddish colors. Their brightest stars are all red giants or the kinds of supergiants that precede the demise of low-mass stars through planetary nebulae. They have low concentrations of the heavy elements, a fact that fits in neatly with the apparent absence of vigorous recycling of matter through the birth and death of stars. And their stars move chaotically, according to no organized scheme.

In their ability to shine brightly in the void, however, ellipticals resemble bulges more closely than they do haloes. I was quickly learning that the haloes of spiral galaxies do not count for much, visually. One certainly did not notice them when gazing at distant spirals—Andromeda, for instance. As I made my first crossing of the Milky Way's halo, I had been misled by the myriad stars that seemed to surround me at all times. Now, on my way out (and much farther away from the disk than I had been when visiting the Magellanic Clouds), I could see the halo for what it was. Compared to the disk or bulge, the sprinkling of bright stars was meager. The halo contained a lot of mass, that much was known from surveys of its gravitational influence, but most of it was very dim, virtually invisible to Earthbound observers and, for that matter, to me. The bulge of the Milky Way, however, shone brightly. It had partaken much more heartily of the carbon, oxygen, and iron produced in the embedded disk, and the reddish cast was less pronounced; in those respects it differed from the ellipticals. But it had the same kind of presence, the concentration of stars, that could make for an imposing galaxy, particularly when these qualities were scaled up to a giant size, as I was soon to discover.

By sheer coincidence, the Milky Way's major companions in the Local Group—the Clouds, Andromeda, and Triangulum—all lay to one side of the disk. Now, having crossed the disk, I saw no such landmarks in my sky as I sped away once more, and I began to feel lonely. The few nearby galaxies I did see were scarcely worth the name—loose, anemic agglomerations of stars rightly called "dwarfs." Just as humans in their original state could feel comfortable only on a rock and water planet like Earth, perhaps spacefaring humans would require the comfort of a robust galaxy to feel secure. As I scanned the desolate space in front of me, I knew that I could not rest until I had found another place where the cycles of change operated vigorously, where stars came and went, and (now that I had seen the Magellanic Clouds) where even whole galaxies exchanged stars and gas and sometimes merged.

I therefore peered out of the Milky Way's halo into greater distances than I had ever thought to travel, to the spaces where all I could see, dimly, were other galaxies, millions of them. The spirals were the most numerous by far, but the places that arrested my attention were marked by huge elliptical galaxies. The most impressive spot on the sky was dominated by at least five big ellipticals, two of which far outshone the other three. They seemed to have herded a thousand or so smaller galaxies into clustering around them, making up a great congregation of galaxies in the direction assigned on Earth to the star pattern Virgo. I knew immediately that this was to be my next destination.

It is not too surprising to find hierarchical arrangements of structure in the Universe. Gravity knows no bounds of scale—no maximum distance beyond which its attraction fails—so, given enough time and opportunity, it will build layer upon layer of structure, *ad infinitum.* What is slightly more surprising is how distinct those levels of structure can be. In the disk of the Milky Way, for example, stars are separated by 10 million times their diameters, and even their planetary systems are separated by distances thousands of times larger than the orbits of the farthest flung planets. (Just look at how long it took humans to bridge those gaps!) But the hierarchy is much less distinct at the level of galaxies, although it seems reasonably secure for the luminous parts—the aggregations of stars, gas, and dust—that were discovered first. Only much later was it found that most of the matter in galaxies is contained in their extensive, invisible haloes. These overlap much more frequently than the luminous parts, linking galaxies to one another physically without much altering the appearance that they are islands in the sky. But my visit to the Magellanic Clouds had shown me the insidious, long-term damage that can be done when galaxies secretly overlap.

As I passed through the outer quarters of the Milky Way, I considered the hidden relationships that might exist among the galaxies I saw spread before me. Many, but not all, seemed to be bound up into clusters. Some of these appeared to be loose aggregates with few members, like the Local Group. Others were groupings as rich in galaxies as the Pleiades cluster is in stars. The galaxies in rich clusters were plainly separated by just a few times their widths; one could only imagine the gravitational struggles that must be going on among their haloes. In a few cases, one didn't have to imagine: Pairs of galaxies were locked in fierce battle, spraying out streamers of stars and gas that contained more matter than the entire Magellanic Cloud system. Where these streamers collided, one sometimes saw the signatures of new stars being born. Why, then, did so many of these tightly grouped galaxies maintain their separate identities, given the forces that worked to mix and homogenize them? The only sensible answer seemed to be lack of time. These clusters must represent tracts of matter still coming together, merely the latest scenes of a galaxy formation opera not yet sung to completion.

Even the clusters of galaxies did not represent the end of the hierarchical sequence. The clusters themselves seemed not to be placed randomly but to be lined up, joined into networks, and linked into grainy membranes that were then curved around great voids.

Thanks to all this structure, I did not have to use dead reckoning to find my way to the Virgo Cluster. I was guided there by one of these metagalactic highways. As I left the Milky Way's halo behind and emerged truly into intergalactic space for the first time, I could appreciate that the Virgo Cluster was not merely an oasis in an otherwise barren desert. I perceived other groups of galaxies nearer by, most of them not much more populous than the Milky Way's own meager Local Group, others containing as many as a hundred or more galaxies. They were not spread uniformly through the space to all sides but, rather, formed a rough sort of corridor that drew me onward. It struck me that, far from being an outpost, the Virgo Cluster was the centerpiece of our corner of the Universe. It was the Local Group that was a way station in Virgo's "supercluster." I may have been leaving home, but I was definitely not leaving the neighborhood.

I suddenly remembered Johannes Kepler and his *Dream*. I had not thought about the *Somnium* since the early years of my trip, but now one of its powerful metaphors stood in startling contrast to my present situation. In his story, the great astronomer had been conveyed to the Moon along a shaft of darkness, the shadow of a lunar eclipse. Far from being a voyage through barren intergalactic space, my way to Virgo was being paved by light—the light of myriad galaxies, each a microcosm of our own. In a little more than 34 years, according to my clocks, I crossed the 60 million light-years and entered the outskirts of the cluster.

31

A City of Galaxies

It was difficult to approach Virgo's cluster of galaxies without feeling the sense of awe that a tourist from "the sticks" feels when visiting a big city for the first time. There was no comparison between the Virgo Cluster and the Milky Way's Local Group. Without counting the lesser dwarf galaxies and other mild concentrations of stars that would have received official titles in the Local Group but would hardly be recognized elsewhere, the Virgo Cluster contained more than a thousand galaxies splashed across a region 10 million light-years on a side. If the Local Group, with its two dozen or so galaxies (of which fewer than 10 made a strong visual impression) represented a village or small town, then the Virgo Cluster would be a good-sized metropolis, with multiple business districts and sprawling suburbs. Close up, it was a maze. The glittering giant ellipticals marked the downtown hubs. The spirals took second place, not because they were anemic, but because the ellipticals were so imposing. At the time, I would have compared it to a Los Angeles or London. Only later did I come to appreciate that even Virgo was far from the acme of its class.

Like the Local Group, Virgo had a handful of galaxies that dominated all the others, any of which would have put the Milky Way or Andromeda to shame. But there was another difference between Virgo and the Local Group that overwhelmed any analogies or distinctions based solely on the scale or arrangement of galaxies: All of the dominant galaxies of Virgo were ellipticals. This much was already apparent from the distant viewpoint of the Milky Way. Whereas a spiral, by virtue of its saucer shape, gives the impression of motion, of slight dynamic imbalance, and perhaps even of tumbling through space (surely an illusion), a dominating elliptical provides a focus, giving a cluster the appearance of stability. Because it is nearly spherical, a dominant elliptical galaxy seems to be stationary and to draw everything else toward it.

The two ellipticals that outshone all the others in Virgo spread out to several times the size of the disk in Andromeda or the Milky Way and seemed to form separate nuclei around which many of the lesser—but still bright—galaxies clustered. Of course, neither of these dominant galaxies or any of the other bright ellipticals could be regarded as the true center of the cluster. Each of them had to be executing an elaborate dance under their mutual gravitational attractions. But their local domains of influence were real. One could carry the urban analogy a step further, identifying neighborhoods within the cluster, each one centered on its local elliptical. These neighborhoods even had distinctive characteristics, with spirals concentrating about one of the elliptical centers, small ellipticals about the other. A third type of galaxy, entirely missing from the Local Group, seemed to be a hybrid that combined a large bulge (or small elliptical) component with a disk that contained stars but little or no gas. These galaxies cast their lots with the small ellipticals.

I felt the cluster's influence long before I reached any of its major concentrations of galaxies. Not the gravity generated by the cluster's enormous mass—I was still moving too fast for that to affect me noticeably—but the cluster's atmosphere. I noticed the onset of its drag because I was especially sensitive to the amount of matter that surrounded my craft, concerned as I was about the fuel supply for my travels. As I am sure you have recognized, it would have been completely impractical for me to carry all the fuel needed for each stage of my journey. During each acceleration phase, most of the fuel would have to be spent near the end, in reaching the maximal Shangri-La factor. But it would have been prohibitive to accelerate all that fuel from rest, not to mention the equal amount of fuel necessary to slow down—*Rocinante* would have been impossibly heavy.

The solution to this problem, of course, is to scoop up matter from interstellar and (lately) intergalactic space as I go along and to use this as my fuel. Fortunately, there is at least *some* matter present, however tenuous, at every known point in the Universe. *Rocinante* had been designed to scoop up enough fuel from the interstellar spaces of the Milky Way to keep me accelerating at 1*g*: This required a catchment area about 10,000 kilometers across, similar to the Earth's radius. A very modest requirement, given the technology of my craft! Plying the Milky Way's halo en route to the Magellanic Clouds had required the scoop radius to be expanded by nearly a factor of 100 to compensate for the much lower density of interstellar matter in these sere regions. But crossing the space between the Local Group and Virgo was a tougher challenge still. There the gas had never been mapped; it was too tenuous to detect, even using the sensitive instruments on board *Rocinante*. Stretching the scoop to several million kilometers across (20 times the distance from Earth to the Moon) was just manageable. Still, I had to monitor the surrounding density continuously to make sure I wouldn't get marooned.

Thus I was relieved when I found myself plunging through a considerably denser atmosphere than I had expected to encounter so far from the heart of the cluster. My X-ray sensors indicated that the gas was 10 times hotter than the gas that filled the Milky Way's halo. Yet, whereas the Milky Way had trouble holding on to its atmosphere (our Galaxy's hot gas was perpetually evaporating away), this even hotter corona seemed to be held in place by gravity. This, my first direct indication of the strength of the Virgo Cluster's gravitational field, was quickly confirmed as I entered the cluster's suburbs and began to measure the speeds of individual galaxies. Each galaxy moved as a unit, under the combined influence of all the other galaxies' attractions, as well as the attraction of any other matter that happened to be present in the cluster. Like the atoms in the hot gas, the galaxies executed random orbits, their speeds regulated so as to keep them from flying away. Still, the speeds were enormous—1000 kilometers per second was typical. Any star moving that fast in the halo of the Milky Way, or any galaxy speeding across the Local Group at such a rate, would have been on a one-way trip to oblivion.

Did it make sense that the gravity in the Virgo Cluster should be so strong? There were a lot of galaxies here, 100 or more for every galaxy in the Local Group, and gravity certainly increased with the amount of matter present. But the cluster was also vastly larger than the size of the Local Group, and it was equally certain that gravity weakened over increasing distances. I had expected the two opposing trends—those due to distance and those due to mass—to cancel each other, yielding a net gravitational effect not too different from that in the Milky Way's vicinity. Yet my measurements told me that gravity was more than 10 times stronger in Virgo.

There was only one sensible explanation. A lot more matter had to be present here than met the eye. My surprise subsided; this was nothing new. The outer reaches of the Milky Way's halo, though invisible, contained more matter than all the visible parts of the Galaxy put together. I remembered how this had been established by remote observations from Earth, long before my departure. In Virgo, the discrepancy between seen and felt mass was much more extreme: There had to be nearly 9 grams of invisible matter for every gram that was visible, and the same trend was repeated in other clusters of galaxies. Were the individual galaxies abnormally heavy, considering their apparent sizes and luminosities? No, it seemed not. One could rule out that possibility by measuring the motions of the stars that made up each galaxy. It seemed instead that most of the invisible matter was not attached to the galaxies at all but formed a collective halo that pervaded the entire cluster, allowing free passage of the galaxies through it. It was comforting to see a pattern repeated at each stage of the hierarchy: Clusters of galaxies had dark haloes, just like each of the galaxies that constituted them. Still, it irked me that I could do no more than confirm what was already known. My presence on the scene gave me no advantage in learning the composition of this elusive matter, let alone capturing a sample of it.

If galaxies could speed through the invisible matter without suffering harm, the same could not be said of their interactions with the hot gas. There was a surprising amount of hot gas, several times more mass than was contained in all the galaxies, and perhaps 10 or 20 percent of the total amount of matter needed to account for the strength of the cluster's gravitational field. I was still 3 million light-years from the nearest dominant elliptical, but the pressure was already higher than it was in the disk of the Milky Way. It was a punishing environment for a galaxy—particularly one flying through it at 1000 kilometers per second—and it took its toll.

I could see the price being paid by galaxies that spent most of their time near the outskirts of the Virgo Cluster but whose orbits took them on quick plunges, at high speed, through the denser regions of the atmosphere. These were mainly spirals, mixed in with raggedy-looking galaxies that reminded me of the Magellanic Clouds. As one galaxy sped by, I observed the damage wreaked by the wind that rushed past as a result of the galaxy's motion through the hot gas. First the galaxy's halo was stripped of its own atmosphere, leaving the disk open to a full assault by the wind. Where the disk's cloud deck was thin or had been punched through, the wind could penetrate right to the disk's midplane, disrupting the coagulation of clouds and, undoubtedly, affecting the normal cycles of star formation. Even where such opportunistic damage was impossible, the wind waged a war of attrition near the disk's outer edge, tearing off shreds of gas constantly. By the time the galaxy had completed its plunge and come out the other side of the cluster, only the complexes of molecular clouds would be intact. The damaged disk, grotesquely truncated, would have less than a billion years to repair itself before the plunge was repeated. I saw another spiral moving so fast that the surrounding gas had no time to get out of the way. It rammed the gas ahead of it, squeezing it into an arc—a shock wave that propagated like a sonic boom through the cluster. A third galaxy, one of the ragged Magellanic types, seemed to be "beside itself" in the sense that nearly its entire complement of hydrogen cloud had been displaced from the disk by the force of the wind and was following behind it, like a ghost.

These effects damaged only the gas of a galaxy: The hot cluster atmosphere was able to flow between the stars with ease. Perhaps this was why, as I moved toward the dominant ellipticals that marked the central zones of the Virgo Cluster, the mix of galaxy types changed, gas-rich spirals and Magellanic types becoming less common and small ellipticals more so. I suspected that spirals simply could not survive in these crowded regions. Where the galaxies—and the hot cluster gas—were most highly concentrated, spirals were scarcest. The ellipticals, on the other hand, seemed to love a crowd. I imagined the fate of a spiral that ventured into one of these crowded zones, its gas stripped away, star formation stymied, perhaps even the stellar component of its disk fading into obscurity. If all that remained was its bulge, could I distinguish it from one of the small ellipticals that now surrounded me? Or perhaps the disk of stars, if not the gas, would remain, and the decimated spiral would take on the appearance of one of the gas-poor, bulge–disk hybrids that also populated these parts. There were other hazards, too, that could have "gotten" the spirals. These regions of concentrated galaxy population presented an increased risk of collisions between galaxies. Surely a spiral's disk would have trouble surviving such an encounter?

By now I had reached the inner sanctum of the Virgo Cluster and had to make a decision: which elliptical to explore. All of the bright ellipticals were old friends from my stargazing days on Earth. They had figured in Charles Messier's list, that undifferentiated eighteenth-century catalogue in which the Crab Nebula featured as number 1, Orion as number 42, the Dumbbell as 27, and Andromeda as 31. The brightest galaxy in Virgo was M49, a star system that could be seen easily with binoculars from Earth, and it was there that I headed first. But my visit was brief, and I left disappointed. I had known, from distant observation, that many spirals were among the galaxies surrounding M49. I had attributed this to the vicissitudes that governed the collection of galaxies into clusters—in my naïve view, this was simply a "neighborhood" favored by spirals. But now I saw the association in a new light. For some reason, this huge elliptical had attracted only a sparse following of galaxies of any kind. Despite its huge mass and luminosity, the gaseous atmosphere that its gravity anchored was meager. Was it any surprise that spirals—a common but fragile type of galaxy—should find this a salubrious environment?

The second brightest galaxy, M87, lay 6 million light-years away. As a destination it looked promising. The combined light of its stars was slightly less than that of M49, but in every other way it reigned over the Virgo Cluster. Galaxies were grouped around it densely, a rich assortment dominated by ellipticals, including several of the other giant ellipticals that made Virgo appear so impressive from afar. The X-ray glare coming from its broad-shouldered halo showed that it had also captured the lion's share of the cluster's gas. But what really riveted my attention was the strange activity going on in its center, a luminous projection that looked like one of the jets I had seen coming out of a newly formed star in Orion, except that it spanned such an enormous distance that I could not grasp how the two phenomena could be related. That settled it: I was going to M87.

32

Brobdingnag

There was something alarming about the size of this galaxy. It was not just its huge diameter, or the fact that it weighed 100 times as much as the Milky Way. It was the way it merged into its surroundings gradually, without ever seeming to end. It was the way its appearance could trick you into thinking its scale was more comprehensible than it really was and then overwhelm you at the last minute. I remember seeing, from a distance, the star-like images that clustered around it and imagining that they were individual giant or supergiant stars. Only as I approached did I realize that each of these "stars" was an entire globular cluster, 100,000 to a million stars each, 10 light-years across. Whereas no more than 200 globulars hugged the inner halo and outer bulge of the Milky Way, M87 had thousands, maybe 10,000, of them spread throughout a region that extended over hundreds of thousands of light-years. And where Andromeda and the Milky Way both had two close companions, M87 was the focus of a swarm of galaxies, some of them comparable in size and brightness to the two great spirals of the Local Group.

M87 had almost certainly boasted even greater numbers of attending galaxies during the course of its history. Many of the galaxies that surrounded it now grazed or even penetrated its immense halo and were in serious jeopardy of being gobbled up. If M87 was trying to conceal a cannibalistic past, its appearance did nothing to allay suspicion. How else could the galaxy have developed its vast, shallow halo of stars, if not by accumulating loose swarms of stars through the destruction of smaller galaxies that got too close. M87 did not look like other elliptical galaxies, which, like the bulges of spirals, seemed to have well-defined boundaries beyond which the light of their stars petered out. For M87, what might have been an entire normal, or even a giant, elliptical galaxy was merely the bright core at its center.

One of the advantages the dominant galaxy in a cluster enjoys is that it suffers little from the blasts of hot gas that can decimate the atmospheres of lesser galaxies. Spirals are most at risk, because their identities are tied so closely to the appearance of their gaseous components. Ellipticals have atmospheres, too, and a fast-moving elliptical on a plunging orbit through the cluster can suffer as much damage as any spiral, though it may not be so obvious to a casual observer. But M87 was immune to such risks. First, with a mass so enormous that it dwarfed that of any nearby galaxy, M87 didn't plunge through the cluster. Rather, it defined the frame of reference against which the plunges of other galaxies were measured. Second and more important, M87 didn't need its own atmosphere—it had appropriated that of the cluster. To a very good approximation, the Virgo Cluster's hot atmosphere was pinned by the gravity of M87 and was carried along with it. There were great eddies of hot gas toward the outer parts of M87's halo, probably stirred up by the motions of nearby galaxies. But the closer I got to the core, the more tightly this gas clung to the galaxy's stolid gravitational field. The swirling motions died away, and I grew accustomed to sailing smoothly through a vast sea of stars.

These calm conditions lasted until I had descended to a point about 300,000 light-years from M87's center. I had sped through more than 100,000 light-years of unchanging atmosphere: constant temperature, constant density, and a monotonous X-ray glow. The concentration of stars around me in the shallow galactic halo had increased so gradually that I had scarcely noticed. But now things began to change. The temperature of the gas, which should have been going up as the gas became compressed under its own weight, began to *decline* as I continued toward the galaxy's center. Small clumps of gas, slightly cooler than the still-hot background, appeared spontaneously and quickly lost all their heat in a blaze of X-ray and ultraviolet radiation. Now cold, these clumps were kneaded and compressed by the pressure of the hotter gas that enveloped them, until they formed dense knots rich in molecules. The remaining hot gas, cooling more gradually than these clumps, began to slump inward toward the center of M87.

Cold clouds peppered the galaxy, too compact to coagulate into large cloud systems and not massive enough, apparently, to form stars. As for the hot gas, it continued to cool and to slide inward gradually, all the time losing ground to the molecular droplets that continually condensed out of it. This would not be a dramatic source of matter for the galaxy—10 Suns' worth of mass per year, even accumulated over 10 billion years, was a pittance for such a large system—but the significance of the flow ran deep. It connected the cluster's atmosphere with the interior of the galaxy in a more vital way than I had expected. I had viewed the hierarchies of scale (star to galaxy to cluster of galaxies, and so on) as an inexorable progression, each one feeding the next. But these were not one-way thoroughfares. I had already learned that the interactions between stars and galaxies ran both ways. Now I learned how deeply the symbiosis was ingrained at the higher levels as well.

Although it was not so thick as to block my view, the spray of dense clouds gave the interstellar space of M87 a variegated texture. I traveled onward, extrapolating what I had seen so far into the regions ahead. The clouds must continue to settle toward the galaxy's nucleus, converging until they had no choice but to coalesce. I pictured familiar terrain, thick cloud decks and star-forming regions, like the Milky Way. But I had forgotten about the dramatic events going on in M87's center.

About 100,000 light-years from the center, I began to encounter bubbles of gas so hot that the particles inside were dashing about at speeds indistinguishable from the speed of light. I had not encountered matter in this form since my visit to the latter-day Crab Nebula, and I immediately suspected that some powerful generator lay below, bristling with electromagnetic fields. In such situations, one has a good chance of seeing radiation emitted by the synchrotron process, the radiation of electrons gyrating under the influence of magnetic forces. This was the mechanism that had been responsible for the garish blue light of the Crab, but here, with the magnetic field much weaker and the hot gaseous atmosphere producing its own glow, only the radio waves generated by the gyrating electrons stood out. To radio-sensitive eyes, I was entering a region of wispy cirrus clouds, their fine strands sketching out the directions of the magnetic lines of force.

At first the ultrahot filaments were juxtaposed with much larger concentrations of (now) lukewarm gas settling in from the cluster. Most of the time I was still immersed in the matter that was slowly cooling and drifting inward. But the closer I got to the center, the more space was taken up by the ultrahot cells. Instead of occupying isolated pockets in the cooler substrate, the bubbles linked up to form interconnected channels, surrounding the regions of cooling gas and squeezing them. This only accentuated their cooling, and it wasn't long before the pockets of lukewarm gas coalesced to form great cool sheets.

About 10,000 or 20,000 light-years out, the ride began to get bumpy. The sky was aglow with radio cirrus. They formed great swirling eddies, reprising the tumult that had characterized M87's outer halo, except that now the turbulence was being driven from below by the activity in the galaxy's nucleus, rather than from above by close encounters with the neighboring galaxies. Intermixed with the cirrus were the sheets of lukewarm gas, also stretched out along the lines of magnetic force. (Cold molecular clumps still peppered the medium sparsely but had never fulfilled their promise of merging to form a dense cloud layer.) A series of sharp jolts informed me that I was passing through a network of shock waves. As though to confirm this, the sheets of cool gas glowed bright, mostly in the pink light of hydrogen, but also with an array of other colors—the reds of nitrogen, the greens of oxygen—that showed the gas was being jostled violently. I passed through one sheet after another. The sheets of cool gas seemed to be lying closer together now, as though they were backing up against an obstacle. And so they were, but it was a curious sort of obstacle: a pressurized cavity so hot and tenuous that it was nearly indistinguishable from a perfect vacuum. I crossed the boundary of the cavity and sailed into open space, a sudden calm surrounding my craft. A shaft of light stretched across the cavity in front of me—I was approaching the jets.

33

Reprise

Despite the pressure of the Virgo Cluster's atmosphere, despite the tremendous gravity of this enormous galaxy, the center of M87 had been cleared of gas out to a distance of 6000 light-years. The jets, which shot outward in opposite directions from the nucleus, had obviously pushed the gas aside. The cavity was not spherical; it was elongated in the same direction as the jets (like the cavity created by the jets of SS 433, I thought). As I scanned the cavity from one end to the other, I could see the impact points where the jets finally slammed into the material they were so effectively holding at bay. I recalled the hot cavity that had welcomed me to the center of the Milky Way. That region had been only 10 or 20 light-years across and could not have been blown open by any sort of emanation from the wimpy central black hole. The Milky Way's cavity had been the product of the hot massive stars that clustered around the black hole, and their winds. Here in M87, the stars took a back seat to a giant black hole, which apparently ruled over the entire central zone of this galaxy.

I had entered the cavity near its middle, halfway between the two impact points, and could see how closely the jets resembled each other. This was no surprise. I had had considerable experience with jets by now, and I assumed that these had emerged from the two sides of a disk that swirled around the black hole. Symmetry is inherent in such an arrangement, and there was no reason why the opposing directions should differ in any fundamental way. What was surprising was the way the jets changed in *appearance* as I guided my craft along the wall of the cavity and headed toward one of the impact points.

Almost immediately, the jet on the opposite side of the cavity from my current position faded, and by the time I had made it two-thirds of the way to the impact point, the opposite jet had effectively become invisible. Meanwhile, the light from the nearby jet had intensified. It shone brightly at all wavelengths from radio waves to X-rays and dominated all other forms of radiation in my vicinity.

This behavior had to be illusory, but the illusion was significant. The matter in these jets was apparently moving at very close to the speed of light, and I was again seeing the famous "headlight effect" that I had read about so often and had witnessed, firsthand, at Cygnus X-1. According to the theory of relativity, a light bulb, traveling at a shade below the speed of light, would not be seen to emit radiation equally in all directions. The light would instead be beamed along its direction of motion, like the headlight of a car. By mapping out the changing asymmetry of the jets' appearance as I moved through the cavity, I estimated the speed of the matter in the jets to differ from that of light by no more than 1 or 2 percent.

Light travels through open space in a straight line, and I would have expected matter traveling at nearly light's speed to do much the same. But to my surprise, the jets were not perfectly straight. In places they jogged and seemed to be a bit unsteady, their directions shifting slightly as I watched. At first I assumed that the jets were being wiggled at their source, but that would imply a certain symmetry in the jogs: Whatever sideways excursion one jet took, the other would take in reverse. However, before I lost sight of the distant jet, I had compared the jets' paths, wiggle for wiggle, and had found an element of randomness that proved they were not merely mimicking one another. What could make such a fast stream change course? The only candidate seemed to be whatever it was that filled the cavity and surrounded the jets. Despite first impressions, the cavity was not a perfect vacuum, and the jets did not treat it as though it were.

I learned this the hard way. Despite its uncanny resemblance to a shaft of light, the jet was made of real gas—just like the jet coming out of a newly formed star or emerging from the disk of the X-ray binary SS 433—and it packed a punch. But it was a very light—that is, a low-density—gas, which is why I had originally ascribed to it (and to the cavity) the properties of a vacuum. When the jet hit the impact point, ramming against the much heavier (that is, denser) atmosphere of the outer galaxy, it was like a high-powered water jet running into a brick wall. The gas of the jet splashed back violently, thus becoming the material that filled the cavity. Thus the jet and the cavity were not so very different in density, pressure, and composition, and it is hardly surprising that they interacted strongly.

The ride first became bumpy as I approached the jet's impact point. This was an unpleasant surprise, because I had gotten used to the smooth conditions that had prevailed nearly the whole way along the wall of the cavity. Before I could reverse course, *Rocinante* was caught up in the full turbulent backwash from the impact and bounced around vigorously. I pulled up short of the impact point itself and headed toward M87's nucleus, skimming along the outside of the jet. This meant that I experienced the same environment the jet was experiencing and was able to follow any events that befell the matter flowing along the jet, though in reverse order. Still close to the impact point, I braced for further turbulence. I could see that the jet was being buffeted harshly and was responding in rather violent jogs from side to side. With all this pounding, it was not surprising that the jet had spread out considerably and was no longer the narrow stream that I had traced outward from the nucleus when I first entered the cavity. The turbulence died down as I moved further away from the impact, only to be replaced by more regular—but still nauseating—thumps. It seems that the jet and the surrounding cavity had conspired to generate periodic impulses of pressure, much as the wind generates periodic swells on the sea. With the jet appearing foreshortened and pointing almost directly at me, I could see that even the smallest shifts in the jet's speed and direction, caused by these pressure waves, could have dramatic effects on its appearance. One after another, a string of bright knots lined up along the jet, each one a shock wave corresponding to a sudden though slight change in the jet's direction. I began to regard it as a minor miracle that the jet stayed as straight as it did.

Several hundred light-years from the nucleus, I began to encounter scraps of cool gas once again. Since entering the cavity, I had encountered only the superheated matter left behind by the jets. Everything else seemed to have been pushed out of the way. After a while this had become a source of worry. I knew that if the jets were being created by the interaction of the black hole's gravity with the mass and motion of surrounding matter, then there had better be more material flowing toward the black hole than just the material left behind by the jets themselves. I therefore looked to these cool filaments as possibly the key to the jets' fuel supply. As I moved toward the center, increasing amounts of this gas came together, swirling with angular momentum, and settled into a disk.

Where did this gas come from? There didn't seem to be any continuous flow from the galaxy's outer atmosphere through the nearly empty cavity. But there were the sheets of cool gas that lay just beyond. Did the cavity occasionally let down its guard, allowing some of this matter to fall in and embrace the hole? Or was this a fossil disk, the relic of some ancient era when matter had last been able to collect unimpeded in the center of M87? When this gas has all been used, will the jets turn off and the black hole go to sleep?

The story told by the gaseous motion was unmistakable. To the first level of approximation, that motion was circular, with a velocity that increased the closer the gas lay to the nucleus. It indicated that a single massive object, located at the exact center of the disk, dominated all other forms of mass that might be distributed throughout the disk—stars, gas clouds, whatever. It confirmed that the mass at the center weighed 3 billion times as much as the Sun. To the second level of approximation, I saw that the gas had found a way to give up its angular momentumso that it might flow inward to feed the black hole. The circular motion was not steady but came in fits and starts. A three-armed pinwheel pattern, outlined in the light of shocked gas—pink, reds, and greens—splayed outward, and the gas, circulating through the spiral, gradually approached the hole.

The spiral-incised platter of swirling gas did not extend all the way to M87's nucleus. Gradually, the gas in the disk became hotter and puffed up into a thick atmosphere enveloping its orbital plane. The pink hydrogen glow excited by the spiral shock waves faded out, as the remaining atoms were dashed to pieces, to be replaced by a glow both harsher and bluer, and the gas itself became much more transparent and harder to see. Flares began to erupt from the turbulent flow, arcing out along paths that clearly traced a strengthening magnetic field. As I closed in, the jet seemed to be disassembling itself into dancing strands of magnetized fluid, which wrapped themselves around a void. Then there was the void! I was facing a black hole once more.

Memories of familiar sights flashed though my brain. Had I been here before? It looked eerily similar to the center of the Milky Way, except that everything here was gigantic. The black hole was scaled up in mass by a factor of 1000; it was 3 billion times heavier than the Sun instead of 3 million. It was swallowing matter at a much more furious rate than the Milky Way's black hole, yet there was something similar about the mode in which it did so. There was no dense, opaque accretion disk here, of the sort I had seen in Cygnus X-1 and SS 433. True, I had seen no jet in the center of the Milky Way, but had it really been absent or just too faint to see? Magnetic flares crackled, here as there, and I could see the loops of magnetic field twisting about the black hole's axis of rotation, leaping upward and eventually joining into the flow that was to become the jet. In the center of the Milky Way, it had looked much the same. Everything had been so tenuous there, but no more so than it was here, near the huge black hole of M87, the environs of which strafed me with the same kind of harsh, transparent glow.

34

On the Brink

All the questions I had repressed, when I stood opposite the Milky Way's great black hole on that first, disappointing leg of my journey, came rushing back into my head. How did this black hole come to rest here? Did it grow from a much smaller seed by swallowing stars and clouds of gas? Could it have started out as the collapsed remnant of a single star of only a few times the mass of the Sun, like the one that had collapsed to become the black hole Cygnus X-1? I remembered daydreaming about a star being torn apart and swallowed by the Milky Way's black hole. Had I waited 10,000 years or so, I really would have seen it happen. But even if the Milky Way's black hole were fed in this way every few thousand years, it is doubtful whether the hole could have grown to its 2.5 million solar masses during the time since the Milky Way formed.

The problem was even more acute in M87. Here, 3 *billion* stars like the Sun would have been required to bring the black hole up to its current mass. The central star cluster contained nowhere near that much matter, and the disk of gas I had seen on my way to the nucleus was dribbling in much too slowly to take up the slack.

Even if the requisite amount of matter were available near the hole, that doesn't mean the hole would grab it. I had seen how many impediments there were to the growth of black holes, the most important being the motion—the angular momentum—of the matter that might be swallowed. The ability of a black hole to grab and swallow stars does not increase very rapidly as the mass of the hole increases, and it starts out shaky when the hole has little mass. If the M87 or Milky Way black hole had started out small, how many stars would have been within its reach? Stars would have had to venture much closer to the hole before being torn apart. Instead of doubling its mass in 5 billion years, say, might it not have taken the hole 10 billion, 15 billion, 100 billion years or more, to swallow enough stars—a time so long that it would have exceeded the age of the Universe, to say nothing of that of the Milky Way or the Virgo Cluster? And even if the black hole could swallow stars at an adequate rate, would the supply of stars have remained adequate over the entire lifetime of the galaxy? As stars were depleted from the danger zone near the black hole, would they have been replaced quickly enough to keep the black hole growing apace?

My natural optimism soon asserted itself. Of course, I had been assuming that the environment at the Milky Way's center or the nucleus of M87 was always similar to its present state. But who was to say that these places had always been so sparse, so relatively gas-free? Hadn't there been a time when the tumbling bar in the Milky Way had not yet assumed such proportions as a gatekeeper against too much inbound gas? During an earlier epoch in the Virgo Cluster, mightn't the flow of the cluster atmosphere into M87 have supplied 100 or 1000 solar masses of gas each year instead of just 10? Perhaps the central black hole, here and in the center of virtually every other galaxy, is merely the dumping ground for much of the debris left over from the galaxy's creation. Or do external events routinely overwhelm the ability of a bar, or a cavity blown by a pair of jets, to keep abundant gas from reaching the black hole. The crucial event might have occurred long after the galaxy formed—the collision of the Milky Way with another galaxy, perhaps, or some hapless (and nameless) spiral torn apart and absorbed by

M87. In either case, the nucleus could have become a dramatic beacon indeed—ultimately as bright as a million or a billion Cygnus X-1's—and the black hole could have grown to its present size in merely a few hundred million years.

I could not be sure that this black hole—or the Milky Way's black hole, for that matter—was homegrown. My visit to Virgo, following on the heels of my stay in the doomed Magellanic Clouds, had shown me that whole galaxies do collide and merge from time to time. If the cluster atmosphere, which, after all, contained much more matter than all of Virgo's galaxies combined, was flowing into M87, then why not spice it up with the occasional nucleus of some small galaxy that happened to contain a black hole. Maybe the featured black hole of M87 or the Milky Way came originally from a galaxy where conditions for growth were more favorable. An alien black hole captured by the Milky Way, like a pebble thrown into a whirlpool, would spin with the swirling disk for a little while but then quickly sink into the center. And if a modest black hole had already been waiting there, then repeated black-hole mergers could have built up the monster I saw, more rapidly than any steady feeding by absorption of gas clouds or stars.

Or perhaps these central black holes antedate their galaxies altogether. Perhaps they are primordial beings from the era before there were any stars at all, when the Universe consisted of lumpy gaseous soup. Could some especially dense clod of this undifferentiated stuff simply have collapsed to form the black hole, and could the galaxy have collected around it later? Which came first, the black hole or the galaxy?

• • •

I could not answer these questions, even after all I had seen, but I was much better prepared to think about them than I had been during the early days of my journey. I no longer believed there were simple answers to any of these questions. Galaxies and their huge, central black holes were probably interrelated in as many complex ways as stars were with galaxies, galaxies with clusters of galaxies, and dark interstellar clouds with glittering young star clusters like the Trapezium. Here, on this vast stage of M87's nucleus, were so many of the same attributes—the same jets; the same swirling motions of the disk, engendered by gravity; the same intermediary action of the magnetic field, stretching and snapping with the transmission of energy from one form to another—that I had seen in so many places before and on so many scales. There was also the black hole, the same engine that drove only modest activities at the center of the Milky Way but here was expanded a thousand times.

The important hierarchy was not just one of objects. It was a symphony of geometric arrangements, patterns of motion, and sequences of events, repeated all over the Universe, over a range of scales that was still difficult for me to comprehend, even though I had seen it firsthand.

I hovered near the brink of M87's huge black hole, wondering what to do next. Did I dare to throw myself into the hole, for science's sake or to resolve my state of uncertainty? The answer in either case was clear: No, out of cowardice if for no other reason. My sickening encounters with tidal forces years ago had left me with a phobia that made it impossible to take such a plunge. Falling into a black hole this large would buy me an hour or two inside before the inevitable stretching forces . . . I could not even bear to think about it. It would be pointless, anyway. There would be no hope of recording what I saw.

Did I dare return to Earth? For me, only a few decades had passed, but the relativity of time, I knew, would preclude a comfortable homecoming. By my reckoning, Earth had aged at least 60 million years since I left. It would be another 60 million years, Earth's time, before I could return to my home planet's vicinity. Did any creatures resembling humans still exist? It seemed unlikely. Was there still a breathable atmosphere? Perhaps my descendants had decamped for other worlds—how was I to find them? I did not know and there was no way to find out. Any signals I could receive from Earth now would be tens of millions of years old. Still, I wagered that the home planet was still there; its lure was almost irresistible. I could not quite let go of the thought of returning.

The alternative? To continue onward, probing other nearby galaxies or venturing more distantly. Another 40 years of travel with the Shangri-La effect could bring me as far as I wanted to go across the Universe—billions of light-years from home if I so chose. Now that I was aware of the great hierarchy of structures, the repetition of themes on ever-widening scales, I began to perceive new possibilities as I stared out into space. There, in the direction that, had I been on Earth, would have framed a beautiful telescopic sight behind the constellation Coma Bernices, was a cluster of galaxies even richer than Virgo's. Six times farther away than Virgo, I could perceive the symmetrical grouping of thousands of galaxies, centered on an elliptical of comparably gargantuan proportions. What could it tell me about the development of cosmic structure that I hadn't already seen? I could see, in my imagination, the swarms of galaxies even greater than those that I had seen in Virgo, merging, blending, smoothing out their structures. The Virgo Cluster and the Local Group might someday come together and perhaps then merge with some even larger cluster of galaxies. Would the Milky Way retain its identity then, or would it have been subsumed into some larger galaxy, just as (by then) it will have swallowed the Magellanic Clouds? Where would this hierarchy of processes, this extension of scales, end? By traveling farther, would I encounter black holes 10 billion and 100 billion times the mass of the Sun, even grander evolutionary cycles, and longer and more powerful jets? I already knew that the answer to at least some of these questions, perhaps all of them, was undoubtedly yes. But I wasn't sure that this knowledge was sufficient justification to spur me onward. The more important question: Would the patterns merely repeat, scaled up or scaled down? Or were there great new organizing principles waiting to be discovered in the galaxies beyond? I wanted to think so, but I wasn't sure.

Certain things I might never be able to explore. Now, I could look out into the far reaches of the Universe, dimly, and see things as they were billions of years ago. I had seen firsthand that the Universe was evolving, in the constant production of heavy elements, the streams of gas falling into or escaping galaxies, the merging and disruption of whole galaxies themselves. The Universe had developed from a very different sort of place. There was an era when the sky shone bright with quasars, huge black holes in their first flush of glory. I can see them out there, their light having been emitted 3, 5, 10 billion years ago. I could be at their locations in 3, 5, 10 billion years from now, but by then they would almost certainly be gone.

Only one choice is easy. The urge to communicate is as strong as ever. I do not know to whom, if anyone, I am addressing this memoir. But not to have recorded my impressions of this voyage and its progress, so far, would have been unthinkable. I therefore cast this memoir into space near the center of this wonderful, enormous galaxy M87, in the hope that it may someday be deciphered.

Acknowledgments

This book owes its existence largely to my fellow astrophysicists, whose collective efforts have led to the images and ideas I try to portray. I have been asked which portions of the story are based on "fact" and which are speculative. This is a difficult question to answer. We astrophysicists hone our cosmic perspectives through constant debate over what is plausible and what is not, given the constraints of observation and logic. Little in astrophysics is ever proven in the sense that a mathematical theorem can be proven. What the narrator presents as facts or observations are my best guesses, based on our current level of understanding. Where an issue is really up in the air, I have made sure that the narrator has not been able to discover the answer, either!

I have tried not to take too many liberties with the principle of causality. Astrophysical objects change over time. I present such objects as SS 433 and the jet in M87 as they are seen today, even though they are likely to be quite different when the narrator reaches them 65,000 and 60 million years, respectively, in the future. I could not ignore this problem in the case of the Crab Nebula, which will have changed beyond recognition by the time the narrator arrives. Therefore, I decided to invent a clone that happens to go off, just in time, a thousand light-years from the old Crab. The chance of this happening in reality is exceedingly small (probably no more than 1 chance in 1000, if that); I hope you will indulge me this artifice.

The narrator's method of travel, which exploits the effect known technically as "relativistic time dilation," is physically sound if not very feasible. It really is possible to travel arbitrarily far across the Universe in a human lifespan (as perceived by the traveler) without violating any laws of nature. I try to be realistic in estimating the fuel and shielding requirements as demanded by physical law, but have little familiarity with the immense literature that exists on methods of space travel. I do not attempt to discuss the exotic sensors that the narrator would surely need to view the scene outside his spacecraft's windows, given the distortions that would result from his motion.

I am grateful to the many friends and colleagues who read and criticized drafts in various stages of completion. I'd like especially to thank Jill Banwell, Caroline Bugler, Annalisa Celotti, Peta Dunstan, Betty Fingold, Michael Nowak, Martin Rees, and Marek Sikora. Most of this book was written while I was on sabbatical at the Institute of Astronomy, University of Cambridge, and the Institute for Theoretical Physics, University of California, Santa Barbara. In addition to the colleagues who provided stimulating environments at these two institutions, I especially thank my landlords and neighbors: Jim and Pat Hennessy in Cambridge, Saral Burdette and David Wieger in Santa Barbara, who did much to make my Earthbound travels so enjoyable. I am also indebted to the John Simon Guggenheim Memorial Foundation and the University of Colorado Council on Research and Creative Work for financial support during my sabbatical.

My editors—Amanda Cook, Sean Abbott, and Connie Day—helped to improve the book immeasurably, and I also thank Jeffrey Robbins for his early enthusiasm and support of the project. Dr. Ka Chun Yu found time to complete his ingenious illustration of *Rocinante*'s path the very week he was preparing to defend his Ph.D. thesis on the Orion Nebula. And my wife, Claire Hay, provided encouragement, sound advice, and thought-provoking discussion throughout.

Finally, I cannot overemphasize the role that public funding plays in making progress in astrophysics possible. Without research support from organizations like the National Science Foundation and the National Aeronautics and Space Administration, there would have been little to write about.

Glossary

accretion disk: disk of gas orbiting in the gravitational field of a body. Internal friction in the gas causes it to spiral toward the body, resulting in accretion.

atomic hydrogen: gas in which hydrogen exists in the form of individual atoms, neither ionized nor paired off into molecules. Much of the interstellar matter in the **Milky Way's** disk takes this form.

Betelgeuse: a **red supergiant** located in the constellation Orion, about 500 light-years from Earth.

black hole: body whose gravitational field is so strong that nothing that falls in, not even light, can escape. Two populations exist: Stellar-mass black holes are formed by the collapse of massive stars; supermassive black holes are of uncertain origin and exist at the centers of most, if not all, galaxies.

bulge: central stellar component of a **spiral galaxy**, consisting of stars on chaotic orbits.

Copernican Principle: guiding principle of astrophysics, according to which no special advantage is accorded to our viewpoint on the Universe. Thus any phenomena we observe are assumed to be commonplace.

Crab Nebula: remnant of a **supernova** explosion observed in A.D. 1054. A **pulsar** that spins 30 times a second powers this compact nebula in Taurus. By the time the narrator reaches it, the nebula has dispersed into interstellar space. M1 in **Messier's** catalogue.

Crab II: an imaginary **supernova** remnant, similar to today's Crab, visited by the narrator 90,000 years after his departure from Earth.

Cygnus X-1: massive X-ray–emitting object in a binary system. Believed to be a **black hole** accreting matter from a disk.

degeneracy pressure: resistance to compression exhibited by dense gases consisting of elementary particles, such as electrons or neutrons, regardless of temperature. Arises from random motions of tightly packed particles predicted by **quantum mechanics.** Prevents **white dwarfs** and **neutron stars** from collapsing.

Dumbbell Nebula: prominent **planetary nebula** located in the constellation Vulpecula. Its shape, resembling two luminous masses connected by a bar, gave rise to its name. M27 in **Messier's** catalogue.

Einstein, Albert (1879–1955): German born, Swiss-American physicist who formulated the **special** and **general theories of relativity.** Also demonstrated the particulate nature of light, a key step in the development of **quantum mechanics.**

elliptical galaxy: roughly spherical galaxy consisting of stars moving on chaotic orbits under their mutual gravitational attractions.

general theory of relativity: theory of gravitation propounded by **Albert Einstein** in 1915, according to which gravity is a manifestation of the curvature of spacetime. Builds on the foundation laid 10 years earlier by Einstein's **special theory of relativity.**

globular cluster: compact, spherical cluster of 100,000 to a few million stars, a few light-years across, orbiting a galaxy. Several hundred globular clusters orbit the **Milky Way**; huge galaxies like **M87** contain thousands.

halo: extended region of stars and gas enveloping a galaxy.

Herschel, William (1738–1822): English astronomer who (among other discoveries) deduced the shape of the **Milky Way**

through star counts, catalogued and classified numerous binary stars, star clusters and **nebulae** (including **planetary nebulae**), and discovered the planet Uranus.

horizon: surface surrounding a **black hole** from within which nothing can escape.

hot star: star with a high surface temperature and blue-white color. Any star that is a few times more massive than the Sun and is still burning hydrogen in its core. Also, an evolved (non-hydrogen burning) star with a surface temperature considerably higher than that of the Sun.

interstellar cloud: region of interstellar space where the density of gas is relatively high and the temperature is relatively low, compared to the surroundings.

interstellar dust: extremely fine granular material that coexists with gas in most regions of interstellar space.

ion: an atom stripped of one or more of its electrons.

jet: fast-moving, narrow stream of **plasma** that shoots out of the center of an **accretion disk** or other rotating system. Associated especially with **black holes** and **protostars.**

Kepler, Johannes (1571–1630): German mathematician and astronomer who deduced three fundamental laws of planetary motion. His *Somnium, sive Astronomia lunaris* ("Dream, or Lunar Astronomy") was written in 1611 and published posthumously in 1634.

light, speed of: a universal constant, 300,000 kilometers per second. A light-year, the distance light travels in a year, equals 9.5 trillion kilometers.

Local Group: a loose grouping of several dozen galaxies that includes the **Milky Way.**

M87: an enormous **elliptical galaxy** in the **Virgo Cluster**—notable for its X-ray–emitting atmosphere, its rich cloud of

globular clusters, and the high-speed **jets** shooting out of its center—that contains a three-billion-Solar-mass **black hole**. Catalogued by **Messier**.

Magellanic Clouds: two prominent satellite galaxies of the **Milky Way**.

Messier, Charles (1730–1817): French astronomer who compiled an early catalogue of astronomical objects, including many prominent **nebulae**, star clusters, and galaxies. Objects in his catalogue are denoted by the prefix M.

Milky Way: large spiral galaxy that contains the Solar System and most destinations described in this book. When capitalized, the word Galaxy refers specifically to the Milky Way.

molecular cloud: relatively cool, dense region of interstellar space in which a large proportion of the atoms have combined to form molecules.

nebula: illuminated patch of interstellar gas.

neutron star: superdense body with a mass similar to that of the Sun but a size of only 10–20 kilometers. Believed to form from the collapsing core of a massive star. Neutron stars have the strongest gravitational fields of all known objects except for **black holes**.

Orion: region of vigorous star formation located about 1500 light-years from Earth. Its appearance from Earth is dominated by the Orion Nebula (M42 in **Messier's** catalogue), which is illuminated by the **Trapezium**. The Orion star-forming region is located in the constellation Orion.

planetary nebula: expanding envelope of gas released by a dying, low-mass star and illuminated by the still-hot stellar core.

plasma: an ionized gas—that is, one in which electrons have been stripped from their atomic nuclei. The hot gases that make up stars, **accretion disks**, and most other systems discussed in this book are plasmas. They are extremely good electrical conductors, which enables them to trap magnetic fields.

protostar: star in the process of formation, consisting of a central core, an **accretion disk,** and, often, **jets**.

pulsar: magnetized, spinning **neutron star** producing beams of radiation that rotate with the star.

quantum mechanics: laws of physics, formulated during the first third of the twentieth century, that describe atoms, molecules, and other small-scale phenomena. This system posits that all forms of energy and matter exhibit characteristics of both particles and waves.

red giant: star that has exhausted the hydrogen in its core and is burning hydrogen in a shell surrounding the core. Characterized by an enormous, cool envelope.

red supergiant: star with an inert core that is burning fuels heavier than hydrogen in shells surrounding the core. Larger and more luminous than a **red giant**.

Shangri-La effect: factor by which time is slowed down for the narrator, relative to other objects in the Galaxy, as a result of the narrator's extremely high speed.

special theory of relativity: Einstein's 1905 theory describing the relationships among space, time, and motion, but neglecting gravity. Gravitational effects are treated in Einstein's **general theory of relativity** (1915).

spiral arms: curved patterns superimposed on the disk of a **spiral galaxy**, corresponding to "traffic jams" in the orbital flow of stars and gas. Associated with the accumulation of large **molecular clouds** and regions of star formation.

spiral galaxy: galaxy containing a prominent disk of stars and gas embedded in an extended **halo**. Gravitational disturbances within the disk give rise to **spiral arms**.

SS 433: binary system notable for producing a pair of **jets** that precess about a fixed axis, like a wobbling top. The jets are believed to arise near the center of an **accretion disk**, but it is not known whether the accreting body is a **neutron star** or a **black hole**.

superbubble: large region of hot interstellar gas created by the combined action of stellar winds and **supernovae** from many **hot stars**.

supernova: enormous explosion of a stellar envelope that occurs when a massive star's core collapses to form a **neutron star**. This term also refers to explosions of **white dwarfs** that have grown too heavy to support themselves.

synchrotron radiation: radiation produced when electrons moving at close to the speed of light gyrate in a magnetic field.

Tarantula Nebula: huge star-forming region in the **Large Magellanic Cloud**. Thousands of times more luminous than Orion. Site of the fictional **supernova** explosion (based on the real Supernova 1987A) witnessed by the narrator.

Texas Symposium: biennial series of conferences, first held in Dallas in 1963, focusing on the latest results in relativistic astrophysics. The meeting site now rotates among international destinations, returning to Texas only occasionally.

tidal force: differential gravitational force that can stretch or squeeze objects.

Trapezium: quartet of young, massive stars responsible for illuminating the Orion Nebula.

Virgo Cluster: cluster of more than a thousand galaxies located about 60 million light-years from the **Milky Way**.

white dwarf: body with a mass similar to that of the Sun but a size similar to that of Earth. Supported against gravity by electron **degeneracy pressure**.

Index